Desktop Audio Technology

J. HOOD

Titles in the Series

Desktop Audio Technology

Digital audio and MIDI principles

Francis Rumsey

AMSTERDAM • BOSTON • HEIDELBERG • LONDON • NEW YORK • OXFORD
PARIS • SAN DIEGO • SAN FRANCISCO • SINGAPORE • SYDNEY • TOKYO

Focal Press is an imprint of Elsevier

Focal Press
An imprint of Elsevier
Linacre House, Jordan Hill, Oxford OX2 8DP
200 Wheeler Road, Burlington MA 01803

First published 2004

British Library Cataloguing in Publication Data
A catalogue record for this book is available from the British Library

Library of Congress Cataloguing in Publication Data
A catalogue record for this book is available from the Library of Congress

ISBN 0 240 51919 1

For information on all Focal Press publications visit our website at:
www.focalpress.com

Typeset by Newgen Imaging Systems (P) Ltd., Chennai, India
Printed and bound in Great Britain by Martins the Printers, Berwick upon Tweed

Contents

Series introduction

The Focal Press Music Technology Series is intended to fill a growing need for authoritative books to support college and university courses in music technology, sound recording, multimedia and their related fields. The books will also be of value to professionals already working in these areas and who want either to update their knowledge or to familiarise themselves with topics that have not been part of their mainstream occupations.

Information technology and digital systems are now widely used in the production of sound and in the composition of music for a wide range of end uses. Those working in these fields need to understand the principles of sound, musical acoustics, sound synthesis, digital audio, video and computer systems. This is a tall order, but people with this breadth of knowledge are increasingly sought after by employers. The series will explain the technology and techniques in a manner which is both readable and factually concise, avoiding the chattiness, informality and technical woolliness of many books on music technology. The authors are all experts in their fields and many come from teaching and research backgrounds.

Dr Francis Rumsey
Series Consultant

1 Introduction to desktop audio technology

1.1 About this book

Audio in computers and other modern desktop devices is inherently digital. This is a book about how digital audio works and how to make best use of its capabilities, including control technologies that are related to the MIDI protocol. The argument for digital audio has been well and truly made by now so there is no particular attempt in this book to justify the merits of it against analog audio. It is a fact that the resources of the computer technology described in this book would not be available unless audio information was converted into a digital form, so the case is closed. The future of audio is now digital, without question, and the devices that audio engineers use are increasingly just generic computing devices that happen to be suitable. Of course audio signals need to be analog at the point where they are converted into acoustic signals, in order for them to be transmitted through the air, but storage, transfer and processing are the topics of this book.

The technology covered in this book is divided into a number of areas. The book is based to some extent on earlier books that are now out of print, bringing together the most important information on digital audio and MIDI in one place. The two are so often combined in applications now that it seems sensible to present them in one book. It also introduces a lot of more recent information on these topics because the field has moved on considerably since those books were written. In this book, therefore, the reader will find coverage of recent developments such as surround sound formats, direct stream digital, new audio project formats, new interfaces and alternatives to MIDI.

The first main chapter, Chapter 2, is concerned with the principles of digital audio and Chapter 3 discusses specific aspects of how this is applied in recording, replay and editing within workstations. Chapter 4 is all about MIDI and synthetic audio control, looking at the means by which artificial sounds can be controlled and manipulated. Chapter 5 deals with hardware of various sorts, including storage devices, buses, computer interfaces and audio processing options. Chapter 6 then concentrates on the question of how to transfer audio between systems, including coverage of audio interfaces, networking and file formats. Chapter 7 deals with audio software or applications, giving examples of different commercial packages that exemplify some of the concepts previously described in practice. The book is not about specific commercial software, however, so readers or manufacturers should look elsewhere if they want detailed coverage of these. Chapter 8 concludes by considering operational issues that may not be familiar to some readers, such as recent spatial reproduction formats, consumer format mastering and quality control issues. It also covers troubleshooting and systems issues such as synchronisation.

Coverage is primarily aimed at professional operations but it is acknowledged that a good definition of this is hard to come by and that many people who use such technology are not professionals. However, the intention is to cover the systems and concepts that apply in operations such as production and post-production, broadcasting and music. To a large extent this book stops at the point where audio leaves the studio environment. Topics such as Internet streaming and consumer delivery of audio have intentionally been avoided as these are large topics in their own right.

1.2 Audio workstations

I have called this book *Desktop Audio Technology* to highlight the coverage of digital audio as it applies in desktop devices such as computer-based audio workstations. There are many different definitions one could use here but for the purposes of this book it is convenient to describe an audio workstation as any computer-based device that stores and processes digital audio and/or control data (such as MIDI data). It is assumed that such devices use some form of direct access storage such as hard disks or solid-state memory, as opposed to tape, as the primary storage medium, so dedicated digital audio tape formats are not covered here.

Many devices that will be termed audio workstations here are also general-purpose multimedia workstations that may handle video and other media data. However the emphasis in this book is on the audio technology and principles involved. There is an increasing use of general purpose computing platforms for audio, both in professional and consumer environments, whereas previously there was wider use of dedicated hardware. This is largely because the processing power available on the average desktop PC is now more than adequate for dealing with multiple channels of audio recording and replay, whereas previously there was a need for hardware that was specially engineered for the purpose. So fast and capacious have desktop machines become that they can now accommodate digital signal processing of audio information (for effects, mixing, and so forth) using the main processor of the computer, at least for a number of channels and depending on the sophistication of the processing required.

The decline in popularity of dedicated audio workstations has had one notable disadvantage to the user – that being the loss of hardware dedicated to control system functions. Modern software packages running on PCs use screen-based interfaces, with keyboard and mouse controls, just like any other application, whereas dedicated devices often had physical controls dedicated to the functions required in editing and processing. This has enabled packages to be sold much more cheaply, but now there is a rapid growth in external physical controllers that restore some of the lost usability of dedicated systems. The result of this is that users can now decide whether they are content to operate a system using screen and mouse or whether they need something larger and more physical.

1.3 Audio and the computer industry

The fact that audio engineers now use general-purpose computers for much of their work highlights the position in which the field of audio engineering now finds itself. Once a clearly distinct field of endeavour with arcane dedicated equipment that needed lining up and careful handling by those in the know, there is a sense (a false one) that anyone can do audio these days. Just give them a computer and a bit of software and 'Bob's your uncle', so to speak.

This situation is not unlike that encountered in the late 1980s by the typesetting, graphic design and publishing industry. Desktop publishing was taking the world by storm and all of a sudden anyone with a computer and a bit of software could produce camera ready artwork. Who needed typesetters or graphic designers any longer? We could do it all ourselves on our desks! Of course it rapidly became clear that there was still a need for people with the creative skills and the time to do it properly, they just had to learn to use the new equipment. If they insisted on sticking with hot metal or pen and paper they ran the risk of being branded as dinosaurs and losing out on a lot of new work. Sales executives, no matter how much they might fancy themselves as designers, are better off selling things and not wasting hours creating second rate brochures, for example.

The same or similar is true of audio, and there is a strong danger that the field will take some backward steps in terms of quality unless audio engineers continue to make clear what is good audio and what will not do. Audio is rapidly being swallowed up by the computer industry and many of the standard architectures and operating systems are incorporating audio features that will dictate what is possible in numerous future applications. Things like correct dithering, sampling frequency conversion, timing issues and so forth, are all crucial to achieving high quality. They are things that have been known about in relation to dedicated audio systems for years but they don't always migrate into the computer industry, which now increasingly thinks it understands audio. So by all means we should take advantage of the economies of scale and the huge benefits that the computer industry brings to audio but we have a duty to ensure that high quality audio remains our key goal.

1.4 Audio and quality

Quality is going both ways in audio at the moment as depicted in Figure 1.1. When digital audio first appeared in consumer and professional forms it was normally fixed at a resolution of 16 bits and a sampling frequency of 44.1 or 48 kHz. This provided very good technical quality that, with good conversion hardware, arguably represented a noticeable improvement over existing analog formats in most respects. Since then there has been development both upwards and downwards in terms of quality.

Quality has gone upwards with the introduction of higher sampling frequencies and resolutions, offering bandwidths into the hundreds of kilohertz if required, and a dynamic range

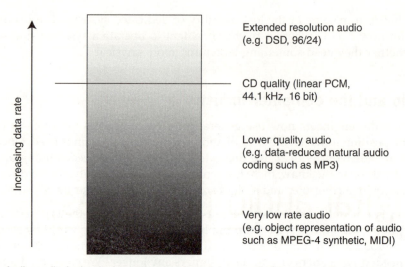

Extended resolution audio
(e.g. DSD, 96/24)

CD quality (linear PCM,
44.1 kHz, 16 bit)

Lower quality audio
(e.g. data-reduced natural audio
coding such as MP3)

Very low rate audio
(e.g. object representation of audio
such as MPEG-4 synthetic, MIDI)

Increasing data rate

Figure 1.1 Audio quality is developing both upwards and downwards from the original reference point of 'CD quality'. Note that lower data rates do not automatically lead to lower quality – this depends on the encoding method used. As the data rate gets lower the method of representation tends towards object representation (description of 'scene' elements, requiring resynthesis by a decoder) rather than natural audio coding (coding of the original audio waveform)

that equals or exceeds that of human hearing. For those audiophiles that still exist in the world this will be appealing, but the economics and practicalities of supplying them with such delights have yet to be completely demonstrated. Spatial quality is also on the increase with the introduction of surround sound and other 3D audio formats. On the other hand, quality is being pushed downwards by the need to deliver audio over potentially very low rate links such as the Internet and mobile communications. Here the question is not how good the quality can be made but how bad it can be made without anyone noticing too much.

Technologies such as MPEG audio coding enable audio to be represented at much lower rates than previously, with minimal impact on audio quality. It is possible to trade off audio quality and bit rate to suit a particular context. Audio quality can also be scaled in MPEG 4 so that a decoder chooses the level of representation depending on the data rate and resources available. There is also a move towards representing audio in the form of objects and control information so that the sound can be resynthesised or rendered in the replay device. This effectively breaks the link between the source and destination in terms of technical quality, because the quality in such a case is dictated primarily by the resources available in the rendering engine, as discussed further in Chapter 8. It is also affected by the completeness of the description of the sound that is supplied. This type of representation is likely to be increasingly common in virtual and synthetic audio authoring environments, for games and the like.

2 Digital audio principles

This chapter explains the fundamental principles of digital audio as they apply in computers. The aim is to aid understanding of the inner workings of equipment so that appropriate operational and technical decisions can be made.

2.1 Analog and digital information

The human senses deal mainly with analog information but computers deal internally with digital information, resulting in a need for conversion between one domain and the other at various points.

Analog information is made up of a continuum of values of some physical quantity, which at any instant may have any value between the limits of the system. For example, a rotating knob may have one of an infinite number of positions – it is therefore an analog controller (see Figure 2.1). A simple switch, on the other hand, can be considered as a digital controller, since it has only two positions – off or on. It cannot take any value in between. The brightness of light that we perceive with our eyes is analog information and as the sun goes down the brightness falls gradually and smoothly, whereas a household light without a dimmer may be either on or off – its state is binary (that is it has only two possible states). A single unit of binary information is called a bit (binary digit) and a bit can only have the value one or zero (corresponding, say, to high and low, or on and off states of the electrical signal).

Electrically, analog information may be represented as a varying voltage or current. If the rotary knob of Figure 2.1 is used to control a variable resistor connected to a voltage supply, its position will affect the output voltage (see Figure 2.2). This, like the knob's position, may occupy any value between the limits – in this case anywhere between 0 V and +V. The switch could be used to control a similar voltage supply and in this case the output voltage could only be either 0 V or +V. In other words the electrical information that resulted would be binary. The high (+V) state could be said to correspond to a binary one and the low state to binary zero (although in many real cases it is actually the other way around). One switch can represent only one binary digit (or bit) but most digital information is made up of more than one bit, allowing digital representations of a number of fixed values.

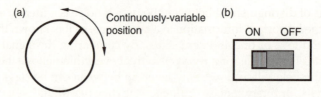

Figure 2.1 (a) A continuously variable control such as a rotary knob is an analog controller. (b) A two-way switch is a digital controller

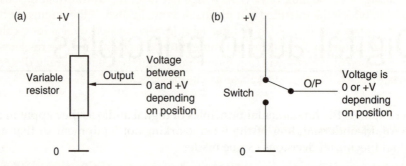

Figure 2.2 Electrical representation of analog and digital information. The rotary controller of Figure 2.1(a) could adjust a variable resistor, producing a voltage anywhere between the limits of 0 and +V, as shown in (a). The switch connected as shown in (b) allows the selection of either 0 or +V states at the output

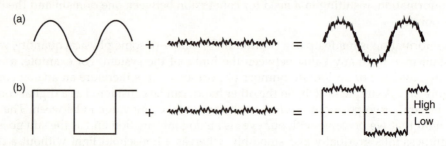

Figure 2.3 When noise is added to an analog signal, as shown at (a), it is not possible for a receiver to know what is the original signal and what is the unwanted noise. With the binary signal, as shown at (b), it is possible to extract the original information even when noise has been added. Everything above the decision level is high and everything below it is low

Analog information in an electrical form can be converted into a digital electrical form using a device known as an analog-to-digital (A/D) convertor. This must be done if the information is to be handled by any logical system such as a computer. The process will be described later. The output of an A/D convertor is a series of binary numerical values representing the analog voltage as accurately as possible, at discrete points in time (sampling instants).

Digital information made up of binary digits is inherently more resilient to noise and interference than analog information, as shown in Figure 2.3. If noise is added to an analog signal it becomes very difficult to tell what is the wanted signal and what is the unwanted noise, as there is no means of distinguishing between the two. If noise is added to a binary signal it is possible to extract the important information at a later stage, as it is known that only two states matter – the high and low, or one and zero states. By comparing the signal amplitude with a fixed decision point it is possible for a receiver to treat everything above the decision point as 'high' and everything below it as 'low'. Any levels in between can be classified in the nearest direction. For any noise or interference to influence the state of a digital signal it must be at least large enough in amplitude to cause a high level to be interpreted as 'low', or vice versa.

The timing of digital signals may also be corrected to some extent, giving digital signals another advantage over analog ones. This arises because digital information has a discrete time structure in which the intended sample instants are known. If the timing of bits in a digital message becomes unstable, such as after having been passed over a long cable with its associated signal distortions, resulting in timing 'jitter', the signal may be re-clocked at a stable rate. There is no equivalent way of removing unwanted speed or timing distortions from analog signals because they have a time-continuous structure.

2.2 Binary number systems

2.2.1 Basic binary

In the decimal number system each digit of a number represents a power of ten. In a binary system each digit or bit represents a power of two (see Figure 2.4). It is possible to calculate the decimal equivalent of a binary integer (whole number) by using the method shown. A number made up of more than one bit is called a binary 'word', and an 8-bit word is called a 'byte' (from 'by eight'). Four bits is called a 'nibble'. The more bits there are in a word the larger the number of states it can represent, with 8 bits allowing 256 (2^8) states and 16 bits allowing 65 536 (2^{16}). The bit with the lowest weight (2^0) is called the least significant bit or LSB and that with the

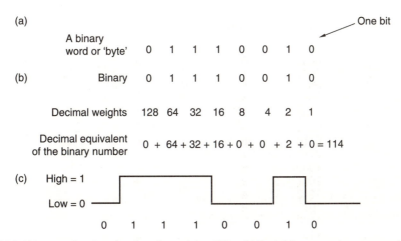

Figure 2.4 (a) A binary number (word or 'byte') consists of bits. (b) Each bit represents a power of two. (c) Binary numbers can be represented electrically in pulse code modulation (PCM) by a string of high and low voltages

greatest weight is called the most significant bit or MSB. The term kilobyte or Kbyte is used to mean 1024 or 2^{10} bytes and the term megabyte or Mbyte represents 1024 Kbytes.

Electrically it is possible to represent a binary word in either serial or parallel form. In serial communication only one connection need be used and the word is clocked out one bit at a time using a device known as a shift register. The shift register is previously loaded with the word in parallel form (see Figure 2.5). The rate at which the serial data is transferred depends on the rate of the clock. In parallel communication each bit of the word is transferred over a separate connection.

Because binary numbers can become fairly unwieldy when they get long, various forms of shorthand are used to make them more manageable. The most common of these is hexadecimal. The hexadecimal system represents decimal values from 0 to 15 using the sixteen symbols 0–9 and A–F, according to Table 2.1. Each hexadecimal digit corresponds to four bits or one

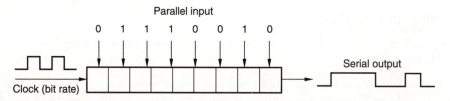

Figure 2.5 A shift register is used to convert a parallel binary word into a serial format. The clock is used to shift the bits one at a time out of the register, and its frequency determines the bit rate. The data may be clocked out of the register either MSB or LSB first, depending on the device and its configuration

Table 2.1 Hexadecimal and decimal equivalents to binary numbers

Binary	Hexadecimal	Decimal
0000	0	0
0001	1	1
0010	2	2
0011	3	3
0100	4	4
0101	5	5
0110	6	6
0111	7	7
1000	8	8
1001	9	9
1010	A	10
1011	B	11
1100	C	12
1101	D	13
1110	E	14
1111	F	15

Figure 2.6 This 16-bit binary number may be represented in hexadecimal as shown, by breaking it up into 4-bit nibbles and representing each nibble as a hex digit

nibble of the binary word. An example showing how a long binary word may be written in hexadecimal (hex) is shown in Figure 2.6 – it is simply a matter of breaking the word up into 4-bit chunks and converting each chunk to hex. Similarly, a hex word can be converted to binary by using the reverse process.

Hexadecimal numbers are often labelled with the prefix '&' to distinguish them from other forms of notation.

2.2.2 Negative numbers

Negative integers are usually represented in a form known as 'twos complement'. Negative values are represented by taking the positive equivalent, inverting all the bits and adding a one. Thus to obtain the 4-bit binary equivalent of decimal minus five (-5^{10}) in binary twos complement form:

$$5^{10} = 0101^2$$

$$-5^{10} = 1010 + 0001 = 1011^2$$

Twos complement numbers have the advantage that the MSB represents the sign (1 = negative, 0 = positive) and that arithmetic may be performed on positive and negative numbers giving the correct result:

e.g. (in decimal):	5
	+ (−3)
	=2
or (in binary):	0101
	+ 1101
	= 0010

The carry bit that may result from adding the two MSBs is ignored.

An example is shown in Figure 2.7 of 4-bit, twos complement numbers arranged in a circular fashion. It will be seen that the binary value changes from all zeros to all ones as it crosses the zero point and that the maximum positive value is 0111 whilst the maximum negative value is 1000, so the values wrap around from maximum positive to maximum negative.

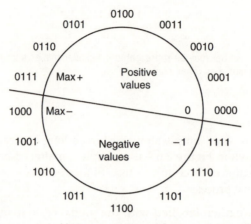

Figure 2.7 Negative numbers represented in twos complement form create a continuum of values where maximum positive wraps round to maximum negative, and bits change from all zeros to all ones at the zero crossing point

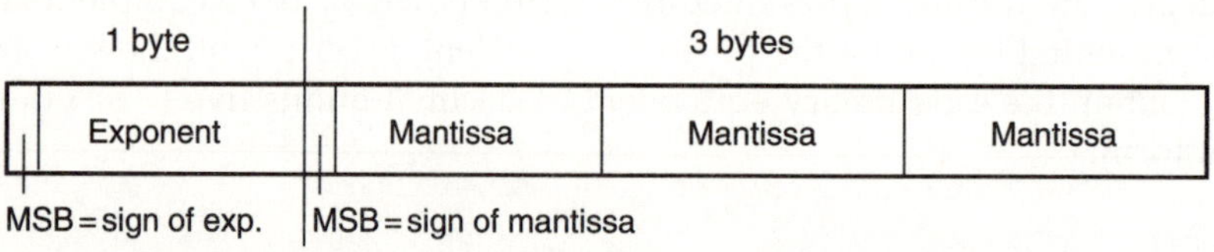

1 byte	3 bytes		
Exponent	Mantissa	Mantissa	Mantissa

MSB = sign of exp. MSB = sign of mantissa

Figure 2.8 An example of floating point number representation in a binary system

2.2.3 Fixed- and floating-point representation

Fixed-point binary numbers are often used in digital audio systems to represent sample values. These are usually integer values represented by a number of bytes (2 bytes for 16 bit samples, 3 bytes for 24 bit samples, etc.). In some applications it is necessary to represent numbers with a very large range, or in a fractional form. Here floating-point representation may be used. A typical floating-point binary number might consist of 32 bits, arranged as 4 bytes, as shown in Figure 2.8. Three bytes are used to represent the mantissa and one byte the exponent (although the choice of number of bits for the exponent and mantissa are open to variance depending on the application). The mantissa is the main part of the numerical value and the exponent determines the power of two to which the mantissa must be raised. The MSB of the exponent is used to represent its sign and the same for the mantissa.

It is normally more straightforward to perform arithmetic processing operations on fixed-point numbers than on floating-point numbers, but signal processing devices are available in both forms.

2.2.4 Logical operations

Most of the apparently complicated processing operations that occur within a computer are actually just a fast sequence of simple logical operations. The apparent power of the

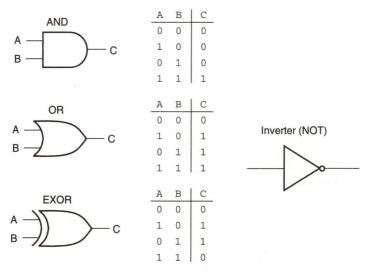

Figure 2.9 Symbols and truth tables for basic logic functions. The inverter shown on the right has an output which is always the opposite of the input. The circle shown on the inverter's output is used to signify inversion on any input or output of a logic gate

computer and its ability to perform complex tasks are really due to the speed with which simple operations are performed.

The basic family of logical operations is shown in Figure 2.9 in the form of a truth table next to the electrical symbol that represents each 'logic gate'. The AND operation gives an output only when both its inputs are true; the OR operation gives an output when either of its inputs are true; and the XOR (exclusive OR) gives an output only when one of its inputs is true. The inverter or NOT gate gives an output which is the opposite of its input and this is often symbolised using a small circle on inputs or outputs of devices to indicate inversion.

2.3 Basic A/D and D/A conversion of control information

In order to convert analog information into digital information it is necessary to measure its amplitude at specific points in time (called 'sampling') and to assign a binary value to each measurement (called 'quantising'). The diagram in Figure 2.10 shows a rotary knob against a fixed scale running from 0 to 9. If one were to quantise the position of the knob it would be necessary to determine which point of the scale it was nearest and unless the pointer was at exactly one of the increments the quantising process would involve a degree of error. It will be seen that the maximum error is actually plus or minus half of an increment, as once the pointer is more than halfway between one increment and the next it should be quantised to the next.

Quantising error is an inevitable side effect in the process of A/D conversion and the degree of error depends on the quantising scale used. Considering binary quantisation, a 4-bit scale offers 16 possible steps, an 8-bit scale offers 256 steps, and a 16-bit scale 65 536. The more bits, the more accurate the process of quantisation.

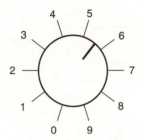

Figure 2.10 A rotary knob's position could be measured against a numbered scale such as the decimal scale shown. Quantising the knob's position would involve deciding which of the limited number of values (0–9) most closely represented the true position

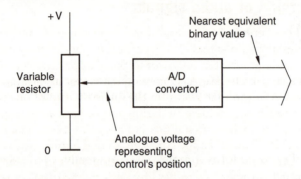

Figure 2.11 In older equipment a control's position was digitised by sampling and quantising an analog voltage derived from a variable resistor connected to the control knob

In older systems, the position of an analog control was first used to derive an analog voltage (as shown earlier in Figure 2.2), then that voltage was converted into a digital value using an A/D convertor (see Figure 2.11). More recent controls may be in the form of binary encoders whose output is immediately digital. Unlike analog controls, switches do not need the services of an A/D convertor for their outputs to be useable by a computer – a switch's output is normally binary in the first place. Only one bit is needed to represent the position of a simple switch.

The rate at which switches and analog controls are sampled depends very much on how important it is that they are updated regularly. Some older audio mixing consoles sampled the positions of automated controls once per television frame (40 ms in Europe), whereas some modern digital mixers sample controls as often as once per audio sample period (roughly 20 μs). Clearly the more regularly a control is sampled the more data will be produced, since there will be one binary value per sample.

Digital-to-analog conversion is the reverse process and involves taking the binary value that represents one sample and converting it back into an electrical voltage. In a control system this voltage could then be used to alter the gain of a voltage-controlled amplifier (VCA), for example, as shown in Figure 2.12. Alternatively it may not be necessary to convert the word back to an analog voltage at all. Many systems are entirely digital and can use the binary value derived from a control's position as a multiplier in a digital signal processing operation. A signal processing operation may be designed to emulate an analog control process.

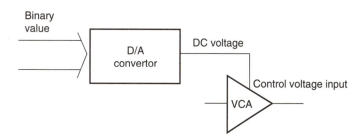

Figure 2.12 A D/A convertor could be used to convert a binary value representing a control's position into an analog voltage. This could then be used to alter the gain of a voltage-controlled amplifier (VCA)

2.4 A/D conversion of audio signals

The process of A/D conversion is of paramount importance in determining the inherent sound quality of a digital audio signal. The technical quality of the audio signal, once converted, can never be made any better, only worse. Some applications deal with audio purely in the digital domain, in which case A/D conversion is not an issue, but most operations involve the acquisition of audio material from the analog world at one time or another. The quality of convertors varies very widely in digital audio workstations and their peripherals because the price range of such workstations is also great. Some stand-alone professional convertors can easily cost as much as the complete digital audio hardware and software for a desktop computer. One can find audio A/D convertors built in to many multimedia desktop computers now, but these are often rather low performance devices when compared with the best available. As will be seen below, the sampling rate and the number of bits per sample are the main determinants of the quality of a digital audio signal, but the design of the convertors determines how closely the sound quality approaches the theoretical limits.

Despite the above, it must be admitted that to the undiscerning ear one 16-bit convertor sounds very much like another and that there is a law of diminishing returns when one compares the increased cost of good convertors with the perceivable improvement in quality. Convertors are very much like wine in this respect.

2.4.1 Audio sampling

An analog audio signal is a time-continuous electrical waveform and the A/D convertor's task is to turn this signal into a time-discrete sequence of binary numbers. The sampling process employed in an A/D convertor involves the measurement or 'sampling' of the amplitude of the audio waveform at regular intervals in time (see Figure 2.13). From this diagram it will be clear that the sample pulses represent the instantaneous amplitudes of the audio signal at each point in time. The samples can be considered as like instantaneous 'still frames' of the audio signal which together and in sequence form a representation of the continuous waveform, rather as the still frames that make up a movie film give the impression of a continuously moving picture when played in quick succession.

In order to represent the fine detail of the signal it is necessary to take a large number of these samples per second. The mathematical sampling theorem proposed by Shannon indicates that at least two samples must be taken per audio cycle if the necessary information about

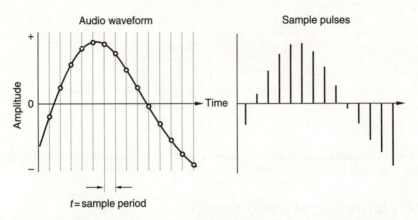

Figure 2.13 An arbitary audio signal is sampled at regular intervals of time *t* to create short sample pulses whose amplitudes represent the instantaneous amplitude of the audio signal at each point in time

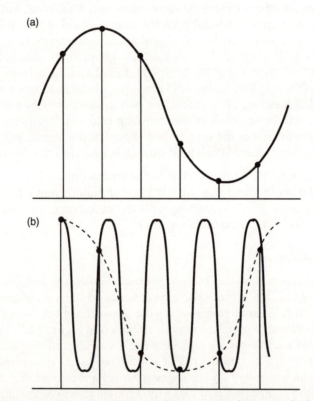

Figure 2.14 In the upper example many samples are taken per cycle of the wave. In the lower example less than two samples are taken per cycle, making it possible for another lower-frequency wave to be reconstructed from the samples. This is one way of viewing the problem of aliasing

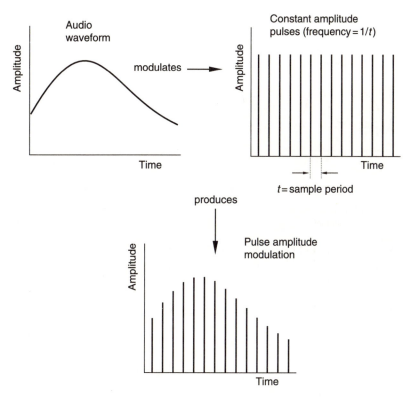

Figure 2.15 In pulse amplitude modulation, the instantaneous amplitude of the sample pulses is modulated by the audio signal amplitude (positive values only shown)

the signal is to be conveyed. It can be seen from Figure 2.14 that if too few samples are taken per cycle of the audio signal then the samples may be interpreted as representing a wave other than that originally sampled. This is one way of understanding the phenomenon known as aliasing. An 'alias' is an unwanted representation of the original signal that arises when the sampled signal is reconstructed during D/A conversion.

Another way of visualising the sampling process is to consider it in terms of modulation, as shown in Figure 2.15. The continuous audio waveform is used to modulate a regular chain of pulses. The frequency of these pulses is the sampling frequency. Before modulation all these pulses have the same amplitude (height), but after modulation the amplitude of the pulses is modified according to the instantaneous amplitude of the audio signal at that point in time. This process is known as pulse amplitude modulation (PAM). The frequency spectrum of the modulated signal is as shown in Figure 2.16. It will be seen that in addition to the 'baseband' audio signal (the original audio spectrum before sampling) there are now a number of additional images of this spectrum, each centred on multiples of the sampling frequency. Sidebands have been produced either side of the sampling frequency and its multiples, as a result of the amplitude modulation, and these extend above and below the sampling frequency and its multiples to the extent of the base bandwidth. In other words these sidebands are pairs of mirror images of the audio baseband.

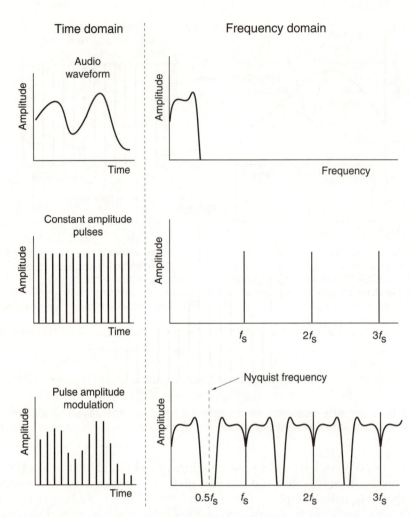

Figure 2.16 The frequency spectrum of a PAM signal consists of a number of repetitions of the audio baseband signal reflected on either side of multiples of the sampling frequency

2.4.2 Filtering and aliasing

It is relatively easy to see why the sampling frequency must be at least twice the highest baseband audio frequency from Figure 2.17. It can be seen that an extension of the baseband above the Nyquist frequency results in the lower sideband of the first spectral repetition overlapping the upper end of the baseband and thus appearing within the audible range that would be reconstructed by a D/A convertor. Two further examples are shown to illustrate the point – the first in which a baseband tone has a low enough frequency for the sampled sidebands to lie above the audio frequency range, and the second in which a much higher frequency tone causes the lower sampled sideband to fall well within the baseband, forming an alias of the original tone that would be perceived as an unwanted component in the reconstructed audio signal.

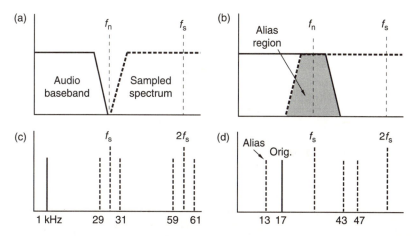

Figure 2.17 Aliasing viewed in the frequency domain. In (a) the audio baseband extends up to half the sampling frequency (the Nyquist frequency f_n) and no aliasing occurs. In (b) the audio baseband extends above the Nyquist frequency and consequently overlaps the lower sideband of the first spectral repetition, giving rise to aliased components in the shaded region. In (c) a tone at 1 kHz is sampled at a sampling frequency of 30 kHz, creating sidebands at 29 and 31 kHz (and at 59 and 61 kHz, etc.). These are well above the normal audio frequency range, and will not be audible. In (d) a tone at 17 kHz is sampled at 30 kHz, putting the first lower sideband at 13 kHz – well within the normal audio range. The 13 kHz sideband is said to be an alias of the original wave

The aliasing phenomenon can be seen in the case of the well-known 'spoked-wheel' effect on films, since moving pictures are also an example of a sampled signal. In film, still pictures (image samples) are normally taken at a rate of 24 per second. If a rotating wheel with a marker on it is filmed it will appear to move round in a forward direction as long as the rate of rotation is much slower than the rate of the still photographs, but as its rotation rate increases it will appear to slow down, stop, and then appear to start moving backwards. The virtual impression of backwards motion gets faster as the rate of rotation of the wheel gets faster and this backwards motion is the aliased result of sampling at too low a rate. Clearly the wheel is not really rotating backwards, it just appears to be. Perhaps ideally one would arrange to filter out moving objects that were rotating faster than half the frame rate of the film, but this is hard to achieve in practice and visible aliasing does not seem to be as annoying subjectively as audible aliasing.

If audio signals are allowed to alias in digital recording one hears the audible equivalent of the backwards-rotating wheel – that is, sound components in the audible spectrum that were not there in the first place, moving downwards in frequency as the original frequency of the signal increases. In basic convertors, therefore, it is necessary to filter the baseband audio signal before the sampling process, as shown in Figure 2.18, so as to remove any components having a frequency higher than half the sampling frequency. It is therefore clear that in practice the choice of sampling frequency governs the high frequency limit of a digital audio system.

In real systems, and because filters are not perfect, the sampling frequency is usually made higher than twice the highest audio frequency to be represented, allowing for the filter to roll off more gently. The filters incorporated into both D/A and A/D convertors have a pronounced effect on sound quality, since they determine the linearity of the frequency response

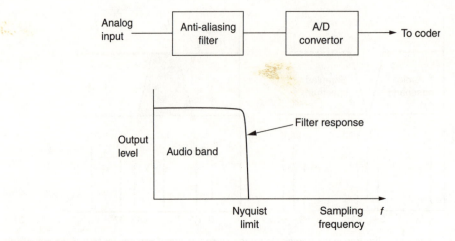

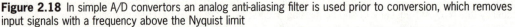

Figure 2.18 In simple A/D convertors an analog anti-aliasing filter is used prior to conversion, which removes input signals with a frequency above the Nyquist limit

within the audio band, the slope with which it rolls off at high frequency and the phase linearity of the system. In a non-oversampling convertor, the filter must reject all signals above half the sampling frequency with an attenuation of at least 80 dB. Steep filters tend to have an erratic phase response at high frequencies and may exhibit 'ringing' due to the high 'Q' of the filter. Steep filters also have the added disadvantage that they are complicated to produce. Although filter effects are unavoidable to some extent, manufacturers have made considerable improvements to analog anti-aliasing and reconstruction filters and these may be retro-fitted to many existing systems with poor filters. A positive effect is normally noticed on sound quality.

The process of oversampling and the use of higher sampling frequencies (see below) has helped to ease the problems of such filtering. Here the first repetition of the baseband is shifted to a much higher frequency, allowing the use of a shallower anti-aliasing filter and consequently fewer audible side effects.

2.4.3 Quantisation

After sampling, the modulated pulse chain is quantised. In quantising a sampled audio signal the range of sample amplitudes is mapped onto a scale of stepped values, as shown in Figure 2.19. The quantiser determines which of a fixed number of quantising intervals (of size Q) each sample lies within and then assigns it a value that represents the mid-point of that interval. This is done in order that each sample amplitude can be represented by a unique binary number in pulse code modulation (PCM) (PCM is the designation for the form of modulation in which signals are represented as a sequence of sampled and quantised binary data words). In linear quantising each quantising step represents an equal increment of signal voltage.

The quantising error magnitude will be a maximum of plus or minus half the amplitude of one quantising step and a greater number of bits per sample will therefore result in a smaller error (see Figure 2.20), provided that the analog voltage range represented remains the same.

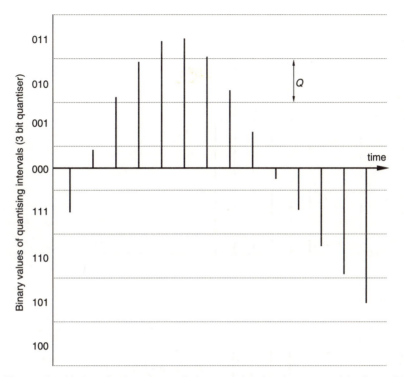

Figure 2.19 When a signal is quantised, each sample is mapped to the closest quantising interval Q, and given the binary value assigned to that interval. (Example of a 3-bit quantiser shown.) On D/A conversion each binary value is assumed to represent the voltage at the mid point of the quantising interval

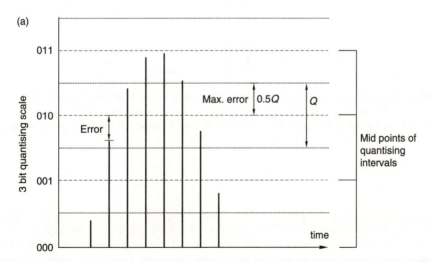

Figure 2.20 In (a) a 3-bit scale is used and only a small number of quantising intervals cover the analog voltage range, making the maximum quantising error quite large. The second sample in this picture will be assigned the value 010, for example, the corresponding voltage of which is somewhat higher than that of the sample.

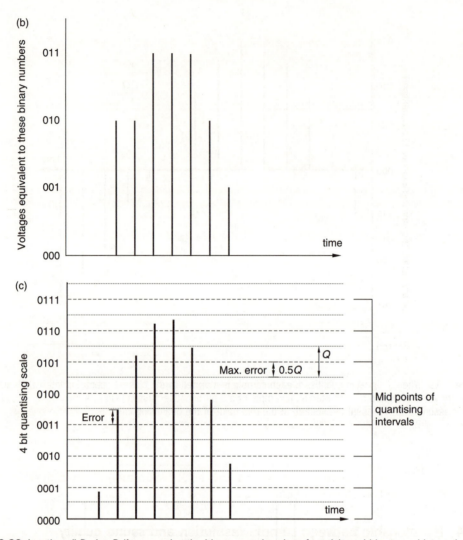

Figure 2.20 (continued) During D/A conversion the binary sample values from (a) would be turned into pulses with the amplitudes shown in (b), where many samples have been forced to the same level owing to quantising. In (c) the 4-bit scale means that a larger number of intervals is used to cover the same range and the quantising error is reduced (expanded positive range only shown for clarity)

Figure 2.21 shows the binary number range covered by digital audio signals at different resolutions using the usual twos complement hexadecimal representation. It will be seen that the maximum positive sample value of a 16-bit signal is &7FFF, whilst the maximum negative value is &8000. The sample value changes from all zeros (&0000) to all ones (&FFFF) as it crosses the zero point. The maximum digital signal level is normally termed

	(a)	(b)	(c)
Max. +ve signal voltage	7F	7FFF	7FFFF
	Positive values		
Zero volts	00	0000	00000
	FF	FFFF	FFFFF
	Negative values		
Max. −ve signal voltage	80	8000	80000

Figure 2.21 Binary number ranges (in hexadecimal) related to analog voltage ranges for different convertor resolutions, assuming twos complement representation of negative values. (a) 8-bit quantiser, (b) 16-bit quantiser, (c) 20-bit quantiser

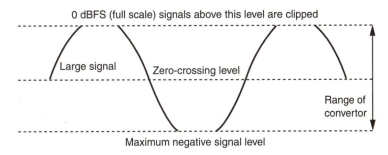

Figure 2.22 Signals exceeding peak level in a digital system are hard-clipped, since no more digits are available to represent the sample value

0 dBFS (FS = full scale). Signals rising above this level are normally hard-clipped, resulting in severe distortion, as shown in Figure 2.22.

2.4.4 Relationship between sample resolution and sound quality

The quantising error may be considered as an unwanted signal added to the wanted signal, as shown in Figure 2.23. Unwanted signals tend to be classified either as distortion or noise, depending on their characteristics, and the nature of the quantising error signal depends very much upon the level and nature of the related audio signal. Here are a few examples, the illustrations for which have been prepared in the digital domain for clarity, using 16-bit sample resolution.

First consider a very low level sine wave signal, sampled then quantised, having a level only just sufficient to turn the least significant bit of the quantiser on and off at its peak (see Figure 2.24(a)). Such a signal would have a quantising error that was periodic, and strongly correlated with the signal, resulting in harmonic distortion. Figure 2.24(b) shows the frequency spectrum, analysed in the digital domain of such a signal, showing clearly the distortion products (predominantly odd harmonics) in addition to the original fundamental. Once the signal

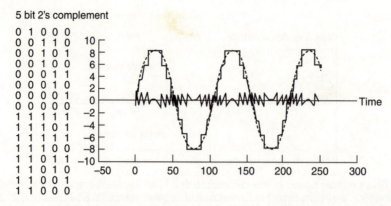

Figure 2.23 Quantising error depicted as an unwanted signal added to the original sample values. Here the error is highly correlated with the signal and will appear as distortion. (Courtesy of Allen Mornington West)

falls below the level at which it just turns on the LSB there is no modulation. The audible result, therefore, of fading such a signal down to silence is that of an increasingly distorted signal suddenly disappearing. A higher-level sine wave signal would cross more quantising intervals and result in more non-zero sample values. As signal level rises the quantising error, still with a maximum value of $\pm0.5Q$, becomes increasingly small as a proportion of the total signal level and the error gradually loses its correlation with the signal.

Consider now a music signal of reasonably high level. Such a signal has widely varying amplitude and spectral characteristics and consequently the quantising error is likely to have a more random nature. In other words it will be more noise-like than distortion-like, hence the term quantising noise that is often used to describe the audible effect of quantising error. An analysis of the power of the quantising error, assuming that it has a noise-like nature, shows that it has an r.m.s. amplitude of $Q/\sqrt{12}$, where Q is the voltage increment represented by one quantising interval. Consequently the signal-to-noise ratio of an ideal n-bit quantised signal can be shown to be:

$6.02n + 1.76$ dB

This implies a theoretical S/N ratio that approximates to just over 6 dB per bit. So a 16-bit convertor might be expected to exhibit a S/N ratio of around 98 dB, and an 8-bit convertor around 50 dB. This assumes an undithered convertor, which is not the normal case, as described below. If a convertor is undithered there will only be quantising noise when a signal is present, but there will be no quiescent noise floor in the absence of a signal. Issues of dynamic range with relation to human hearing are discussed further in Section 2.6.

2.4.5 Use of dither

The use of dither in A/D conversion, as well as in conversion between one sample resolution and another, is now widely accepted as correct. It has the effect of linearising a normal

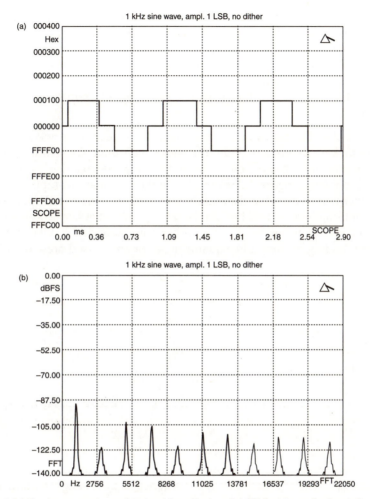

Figure 2.24 (a) A 1 kHz sine wave at very low level (amplitude ±1 LSB) just turns the least significant bit of the quantiser on and off. Analysed in the digital domain with sample values shown in hex on the vertical axis and time in ms on the horizontal axis. (b) Frequency spectrum of this quantised sine wave, showing distortion products

convertor (in other words it effectively makes each quantising interval the same size) and turns quantising distortion into a random, noise-like signal at all times. This is desirable for a number of reasons. Firstly because white noise at very low level is less subjectively annoying than distortion; secondly because it allows signals to be faded smoothly down without the sudden disappearance noted above; and thirdly because it often allows signals to be reconstructed even when their level is below the noise floor of the system. Undithered audio signals begin to sound 'grainy' and distorted as the signal level falls. Quiescent hiss will disappear if dither is switched off, making a system seem quieter, but a small amount of continuous hiss is considered preferable to low level distortion. The resolution of modern high resolution convertors is such that the noise floor is normally inaudible in any case.

Dithering a convertor involves the addition of a very low level signal to the audio whose amplitude depends upon the type of dither employed (see below). The dither signal is usually noise, but may also be a waveform at half the sampling frequency or a combination of the two. A signal that has not been correctly dithered during the A/D conversion process cannot thereafter be dithered with the same effect, because the signal will have been irrevocably distorted. How then does dither perform the seemingly remarkable task of removing quantising distortion?

It was stated above that the distortion was a result of the correlation between the signal and the quantising error, making the error periodic and subjectively annoying. Adding noise, which is a random signal, to the audio has the effect of randomising the quantising error and making it noise-like as well (shown in Figure 2.25(a) and (b)). If the noise has an amplitude similar in level to the LSB (in other words, one quantising step) then a signal lying exactly at the decision point between one quantising interval and the next may be quantised either upwards or downwards, depending on the instantaneous level of the dither noise added to it. Over time this random effect is averaged, leading to a noise-like quantising error and a fixed noise floor in the system.

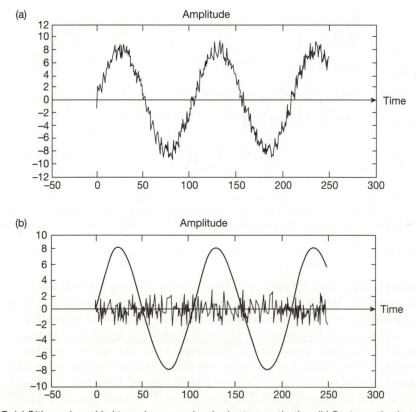

Figure 2.25 (a) Dither noise added to a sine wave signal prior to quantisation. (b) Post-quantisation the error signal is now random and noise-like. (Courtesy of Allen Mornington West)

Figure 2.26(a) shows the same low-level sine wave as in Figure 2.24, but this time with dither noise added. The quantised signal retains the cyclical pattern of the 1 kHz sine wave but is now modulated much more frequently between states, and a random element has been added. The frequency spectrum of this signal, Figure 2.26(b), shows a single sine wave component accompanied by a flat noise floor. Figure 2.26(c) and (d) show the waveform and

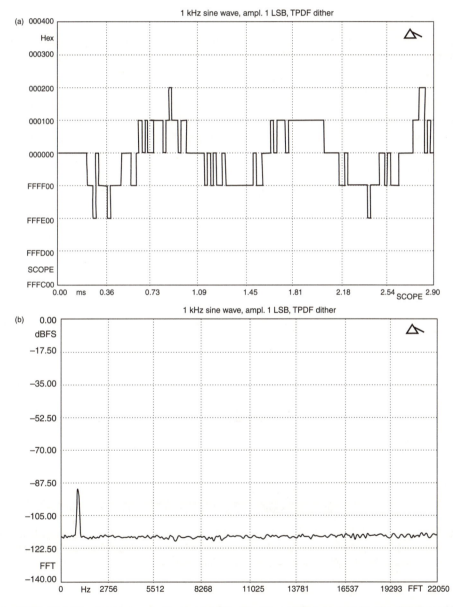

Figure 2.26 (a) 1 kHz sine wave, amplitude ±1 LSB, with dither added, analysed in the digital domain. (b) Spectrum of this dithered low level sine wave showing lack of distortion and flat noise floor.

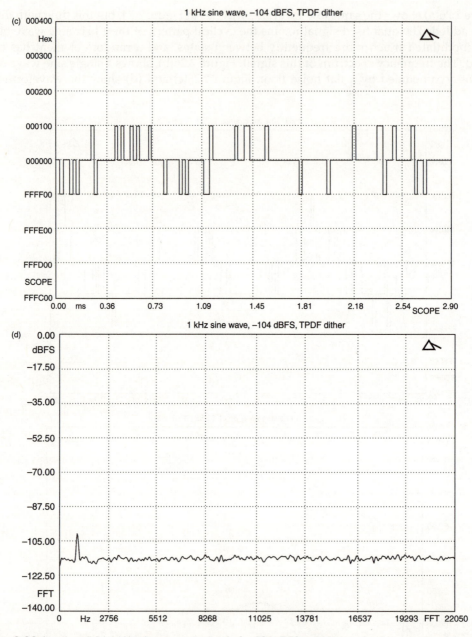

Figure 2.26 (continued) (c) 1 kHz sine wave at a level of −104 dBFS with dither, showing occasional modulation of LSB. (d) Spectrum of this signal showing that it is still possible to discern the original signal. An undithered 16-bit system would be incapable of representing a signal below about −97 dBFS

spectrum of a dithered sine wave at a level that would be impossible to represent in an undithered 16-bit system. In this case the LSB is in the zero state much more frequently than the one state, but an element of the original 1 kHz period can still be seen in its modulation pattern if studied carefully. The duty cycle of the LSB modulation (ratio between time on and time off) varies with the instantaneous amplitude of the original signal. When this is passed through a D/A convertor and reconstruction filter the result is a pure sine wave signal plus noise, as can be seen from the spectrum analysis.

Dither is also used in digital processing devices such as mixers, but in such cases it is introduced in the digital domain as a random number sequence (the digital equivalent of white noise). In this context it is used to remove low-level distortion in signals whose gains have been altered and to optimise the conversion from high resolution to lower resolution during post-production (see below).

2.4.6 Types of dither

Research has shown that certain types of dither signal are more suitable than others for high quality audio work. Dither noise is often characterised in terms of its probability distribution, which is a statistical method of showing the likelihood of the signal having a certain amplitude. A simple graph such as that shown in Figure 2.27 is used to indicate the shape of the distribution. The probability is the vertical axis and the amplitude in terms of quantising steps is the horizontal axis.

Logical probability distributions can be understood simply by thinking of the way in which dice fall when thrown (see Figure 2.28). A single die throw has a rectangular probability distribution function (RPDF), because there is an equal chance of the throw being between 1 and 6 (unless the die is weighted!). The total value of a pair of dice, on the other hand, has a roughly triangular probability distribution function (TPDF) with the peak grouped on values from 6 to 8, because there are more combinations that make these totals than there are combinations making 2 or 12. Going back to digital electronics, one could liken the dice to random number generators and see that RPDF dither could be created using a single random number generator, and that TPDF dither could be created by adding the outputs of two RPDF generators.

RPDF dither has equal likelihood that the amplitude of the noise will fall anywhere between zero and maximum, whereas TPDF dither has greater likelihood that the amplitude will be

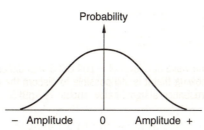

Figure 2.27 A probability distribution curve for dither shows the likelihood of the dither signal having a certain amplitude, averaged over a long time period

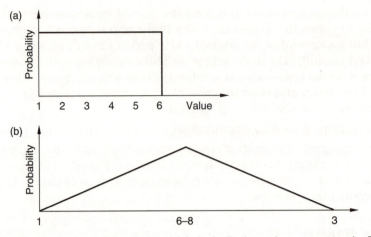

Figure 2.28 Probability distributions of dice throws. (a) A single die throw shows a rectangular PDF. (b) A pair of thrown dice added together has a roughly triangular PDF (in fact it is stepped)

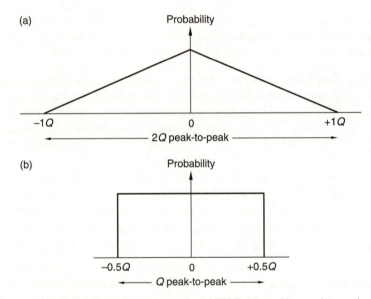

Figure 2.29 Most suitable digital dither signals for audio. (a) TPDF dither with a peak-to-peak amplitude of $2Q$. (b) RPDF dither with an amplitude of $1Q$

zero than that it will be maximum. Analog white noise has Gaussian probability, whose shape is slightly more unusual than either of the logically generated dithers. Although RPDF, TPDF and Gaussian dither can have the effect of linearising conversion and removing distortion, RPDF dither tends to result in noise modulation at low signal levels. The most suitable dither noise is found to be TPDF with a peak-to-peak amplitude of $2Q$ (see Figure 2.29). If RPDF dither is used it should have a peak-to-peak amplitude of $1Q$.

Whilst it is easy to generate ideal logical PDFs in the digital domain, it is likely that the noise source present in many convertors will be analog and therefore Gaussian in nature. With Gaussian noise, the optimum r.m.s. amplitude for the dither signal is 0.5Q, at which level noise modulation is minimised but not altogether absent. Dither at this level has the effect of reducing the undithered dynamic range by about 6 dB, making the dithered dynamic range of an ideal 16 bit convertor around 92 dB.

2.4.7 Oversampling in A/D conversion

Oversampling involves sampling audio at a higher frequency than strictly necessary to satisfy the Nyquist criterion. Normally, though, this high rate is reduced to a lower rate in a subsequent digital filtering process, in order that no more storage space is required than for conventionally sampled audio. It works by trading off sample resolution against sampling rate, based on the principle that the information carrying capacity of a channel is related to the product of these two factors. Samples at a high rate with low resolution can be converted into samples at a lower rate with higher resolution, with no overall loss of information (this is related to sound quality). Oversampling has now become so popular that it is the norm in most high-quality audio convertors.

Although oversampling A/D convertors often quote very high sampling rates of up to 128 times the basic rates of 44.1 or 48 kHz, the actual rate at the digital output of the convertor is reduced to a basic rate or a small multiple thereof (e.g. 48, 96 or 128 kHz). Samples acquired at the high rate are quantised to only a few bits resolution and then digitally filtered to reduce the sampling rate, as shown in Figure 2.30. The digital low-pass filter limits the bandwidth of the signal to half the basic sampling frequency in order to avoid aliasing, and this is coupled with 'decimation'. Decimation reduces the sampling rate by dropping samples from the oversampled stream. A result of the low-pass filtering operation is to increase the word length of the samples very considerably. This is not simply an arbitrary extension of the wordlength, but an accurate calculation of the correct value of each sample, based on the values of surrounding samples (see Section 2.11 on digital signal processing). Although oversampling convertors quantise samples initially at a low resolution, the output of the decimator consists of samples at a lower rate with more bits of resolution. The sample resolution can then be shortened as necessary (see Section 2.8 on requantising) to produce the desired word length.

Oversampling brings with it a number of benefits and is the key to improved sound quality at both the A/D and D/A ends of a system. Because the initial sampling rate is well above

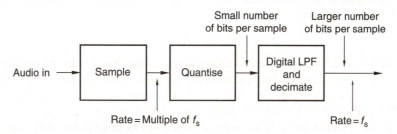

Figure 2.30 Block diagram of oversampling A/D conversion process

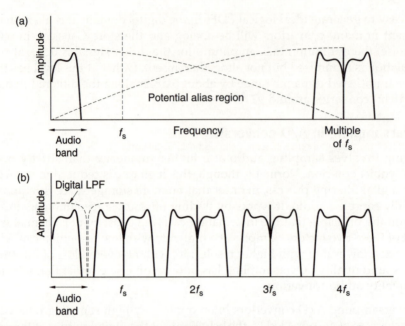

Figure 2.31 (a) Oversampling in A/D conversion initially creates spectral repetitions that lie a long way from the top of the audio baseband. The dotted line shows the theoretical extension of the baseband and the potential for aliasing, but the audio signal only occupies the bottom part of this band. (b) Decimation and digital low pass filtering limits the baseband to half the sampling frequency, thereby eliminating any aliasing effects, and creates a conventional collection of spectral repetitions at multiples of the sampling frequency

the audio range (often tens or hundreds of times the nominal rate) the spectral repetitions resulting from PAM are a long way from the upper end of the audio band (see Figure 2.31). The analog anti-aliasing filter used in conventional convertors is replaced by a digital decimation filter. Such filters can be made to have a linear phase response if required, resulting in higher sound quality. If oversampling is also used in D/A conversion the analog reconstruction filter can have a shallower roll-off. This can have the effect of improving phase linearity within the audio band, which is known to improve audio quality. In oversampled D/A conversion, basic rate audio is up-sampled to a higher rate before conversion and reconstruction filtering. Oversampling also makes it possible to introduce so-called 'noise shaping' into the conversion process, which allows quantising noise to be shifted out of the most audible parts of the spectrum.

Oversampling *without* subsequent decimation is a fundamental principle of Sony's Direct Stream Digital system, described in Section 2.7.

2.4.8 Noise shaping in A/D conversion

Noise shaping is a means by which noise within the most audible parts of the audio frequency range is reduced at the expense of increased noise at other frequencies, using a process that

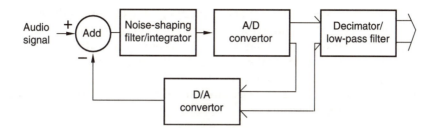

Figure 2.32 Block diagram of a noise shaping delta-sigma A/D convertor

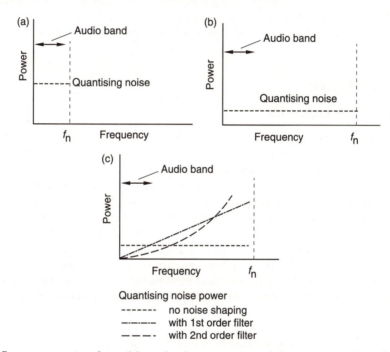

Figure 2.33 Frequency spectra of quantising noise. In a non-oversampled convertor, as shown in (a), the quantising noise is constrained to lie within the audio band. In an oversampling convertor, as shown in (b), the quantising noise power is spread over a much wider range, thus reducing its energy in the audio band. (c) With noise shaping the noise power within the audio band is reduced still further, at the expense of increased noise outside that band

'shapes' the spectral energy of the quantising noise. It is possible because of the high sampling rates used in oversampling convertors. A high sampling rate extends the frequency range over which quantising noise is spread, putting much of it outside the audio band.

Quantising noise energy extends over the whole baseband, up to the Nyquist frequency. Oversampling spreads the quantising noise energy over a wider spectrum, because in over-sampled convertors the Nyquist frequency is well above the upper limit of the audio band. This has the effect of reducing the in-band noise by around 3 dB per octave of oversampling (in other words, a system oversampling at twice the Nyquist rate would see the noise power within the audio band reduced by 3 dB).

In oversampled noise-shaping A/D conversion an integrator (low-pass filter) is introduced before the quantiser, and a D/A convertor is incorporated into a negative feedback loop, as shown in Figure 2.32. This is the so-called 'sigma-delta convertor'. Without going too deeply into the principles of such convertors, the result is that the quantising noise (introduced after the integrator) is given a rising frequency response at the input to the decimator, whilst the input signal is passed with a flat response. There are clear parallels between such a circuit and analog negative-feedback circuits.

Without noise shaping, the energy spectrum of quantising noise is flat up to the Nyquist frequency, but with first-order noise shaping this energy spectrum is made non-flat, as shown in Figure 2.33. With second-order noise shaping the in-band reduction in noise is even greater, such that the in-band noise is well below that achieved without noise shaping.

2.5 D/A conversion

2.5.1 A basic D/A convertor

The basic D/A conversion process is shown in Figure 2.34. Audio sample words are converted back into a staircase-like chain of voltage levels corresponding to the sample values. This is achieved in simple convertors by using the states of bits to turn current sources on or off, making up the required pulse amplitude by the combination of outputs of each of these sources. This staircase is then 'resampled' to reduce the width of the pulses before they are passed through a low-pass reconstruction filter whose cut-off frequency is half the sampling frequency. The effect of the reconstruction filter is to join up the sample points to make a smooth waveform. Resampling is necessary because otherwise the averaging effect of the filter would result in a reduction in the amplitude of high-frequency audio signals (the so-called 'aperture effect'). Aperture effect may be reduced by limiting the width of the sample pulses to perhaps one-eighth of the sample period. Equalisation may be required to correct for aperture effect.

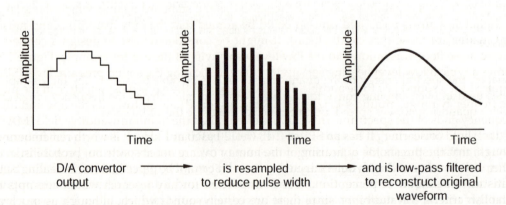

Figure 2.34 Processes involved in D/A conversion (positive sample values only shown)

2.5.2 Oversampling in D/A conversion

Oversampling may be used in D/A conversion, as well as in A/D conversion. In the D/A case additional samples must be created in between the Nyquist rate samples in order that conversion can be performed at a higher sampling rate. These are produced by sample rate conversion of the PCM data. (Sample rate conversion is introduced in Section 2.11.7.) These samples are then converted back to analog at the higher rate, again avoiding the need for steep analog filters. Noise shaping may also be introduced at the D/A stage, depending on the design of the convertor, to reduce the subjective level of the noise.

A number of advanced D/A convertor designs exist which involve oversampling at a high rate, creating samples with only a few bits of resolution. The extreme version of this approach involves very high rate conversion of single bit samples (so-called 'bit stream conversion'), with noise shaping to optimise the noise spectrum of the signal. The theory of these convertors is outside the scope of this book.

2.6 Sound quality versus sample rates and resolutions

The question often arises as to what sample rate and resolution is necessary for a certain quality of audio. What are the effects of selecting certain values? Are there standards? This section aims to provide some guidelines in this area, with reference to the capabilities of human hearing that must be considered the ultimate arbiter in this matter.

2.6.1 Psychoacoustic limitations

It is possible with digital audio to approach the limits of human hearing in terms of sound quality. In other words, the unwanted artefacts of the process can be controlled so as to be close to or below the thresholds of perception. It is also true, though, that badly engineered digital audio can sound poor and that the term 'digital' does not automatically imply high quality. The choice of sampling parameters and noise shaping methods, as well as more subtle aspects of convertor design, affect the frequency response, distortion and perceived dynamic range of digital audio signals.

The human ear's capabilities should be regarded as the standard against which the quality of digital systems is measured, since it could be argued that the only distortions and noises that matter are those that can be heard. It might be considered wise to design a convertor whose noise floor was tailored to the low level sensitivity of the ear, for example. Figure 2.35 shows a typical low level hearing sensitivity curve, indicating the sound pressure level (SPL) required for a sound just to be audible. It will be seen that the ear is most sensitive in the middle frequency range, around 4 kHz, and that the response tails off towards the low and high frequency ends of the spectrum. This curve is often called the 'minimum audible field (MAF)' or 'threshold of hearing'. It has an SPL of 0 dB (ref. 20 µPa) at 1 kHz. It is worth remembering, though, that the thresholds of hearing of the human ear are not absolute but probabilistic. In other words, when trying to determine what can and cannot be perceived one is dealing with statistical likelihood of perception. This is important for any research which attempts to establish criteria for audibility, since there are certain sounds which, although as much as 10 dB below the accepted thresholds, have a statistical likelihood of perception which may

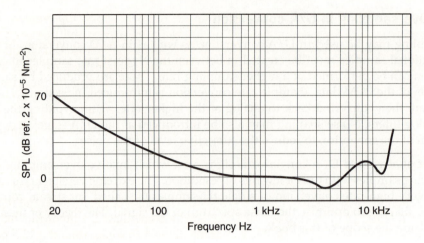

Figure 2.35 Hearing threshold curve

approach certainty in some cases. Also, some listeners are known to be more sensitive than others.

Dynamic range could be said to be equal to the range between the MAF and the loudest sound tolerable. The loudest sound tolerable depends very much on the person, but the threshold of 'pain' is usually said to occur between 130 and 140 dB SPL. The absolute maximum dynamic range of human hearing is therefore around 140 dB at 1 kHz, but quite a lot less than that at low and high frequencies. Whether or not it is desirable to be able to record and reproduce such a wide dynamic range is debatable.

Work carried out by Louis Fielder and Elizabeth Cohen attempted to establish the dynamic range requirements for high quality digital audio systems by investigating the extremes of sound pressure available from acoustic sources and comparing these with the perceivable noise floors in real acoustic environments. Using psychoacoustic theory, Fielder was able to show what was likely to be heard at different frequencies in terms of noise and distortion, and where the limiting elements might be in a typical recording chain. Having defined dynamic range as 'the ratio between the r.m.s. maximum undistorted sine wave level producing peak levels equal to a particular peak level and the r.m.s. level of 20 kHz band-limited white noise that has the same apparent loudness as a particular audio chain's equipment noise in the absence of a signal', he proceeded to show that the just audible level of a 20 kHz bandwidth noise signal was about 4 dB SPL, and that a number of musical performances reached levels of between 120 and 129 dB SPL in favoured listening positions. From this he determined a dynamic range requirement of 122 dB for natural reproduction. Taking into account microphone performance and the limitations of consumer loudspeakers, this requirement dropped to 115 dB for consumer systems.

2.6.2 Sampling rate

The choice of sampling rate determines the maximum audio bandwidth available. There is a strong argument for choosing a sampling rate no higher than is strictly necessary, in other words not much higher than twice the highest audio frequency to be represented. This often starts arguments over what is the highest useful audio frequency and this is an area over which heated debates have raged. Conventional wisdom has it that the audio frequency band extends up to 20 kHz, implying the need for a sampling frequency of just over 40 kHz for high quality audio work. There are in fact two standard sampling frequencies between 40 and 50 kHz: the compact disc rate of 44.1 kHz and the so-called 'professional' rate of 48 kHz. These are both allowed in the original AES5 standard of 1984, which sets down preferred sampling frequencies for digital audio equipment. Table 2.2 is an attempt to summarise the variety of sampling frequencies in existence and their applications.

The 48 kHz rate was originally included because it left a certain amount of leeway for downward varispeed in tape recorders. When many digital recorders are varispeeded, their sampling rate changes proportionately and the result is a shifting of the first spectral repetition of the audio baseband. If the sampling rate is reduced too far, aliased components may become audible. Most professional digital tape recorders allowed for only around ±12.5 per cent of varispeed for this reason. It is possible now, though, to avoid such problems using digital low

Table 2.2 Common sampling rates encountered in digital audio applications

Frequency (kHz)	Application
8	Telephony (speech quality). ITU-T G711 standard.
16	Used in some telephony applications. ITU-T G722 data reduction.
18.9	CD-ROM/XA and CD-I standard for low–moderate quality audio using ADPCM to extend playing time.
~22.05	Half the CD frequency is 22.05 kHz. Used in some moderate quality computer applications. The original Apple Macintosh audio sampling frequency was 22254.5454... Hz.
32	Used in some broadcast coding systems, e.g. NICAM. DAT long play mode. AES 5 secondary rate.
37.8	CD-ROM/XA and CD-I standard for intermediate quality audio using ADPCM.
44.056	A slight modification of the 44.1 kHz frequency used in some older equipment to synchronise digital audio with the NTSC television frame rate of 29.97 frames per second. Such 'pull-down' rates are sometimes still encountered in video sync situations.
44.1	CD sampling frequency. AES 5 secondary rate.
47.952	Occasionally encountered when 48 kHz equipment is used in NTSC video operations. Another 'pull-down' rate, ideally to be avoided.
48	AES 5 primary rate for professional applications.
88.2	Twice the CD sampling frequency. Optional for DVD-Audio.
96	AES 5-1998 secondary rate for high bandwidth applications. Optional for DVD-Video and DVD-Audio.
176.4 and 192	Four times the basic standard rates, optional in DVD-Audio.
2.8224 MHz	DSD sampling frequency.

pass filters whose cut-off frequency varies with the sampling frequency, or by using digital signal processing to vary the pitch of audio without varying the output sampling frequency.

The 44.1 kHz frequency had been established earlier on for the consumer compact disc and is very widely used in the industry. In fact in many ways it has become the sampling rate of choice for most professional recordings. It allows for full use of the 20 kHz audio band and oversampling convertors allow for the use of shallow analog anti-aliasing filters which avoid phase problems at high audio frequencies. It also generates 10 per cent less data per second than the 48 kHz rate, making it economical from a storage point of view.

A rate of 32 kHz is used in some broadcasting applications, such as NICAM 728 stereo TV transmissions, and in some radio distribution systems. Television and FM radio sound bandwidth is limited to 15 kHz and a considerable economy of transmission bandwidth is achieved by the use of this lower sampling rate. The majority of important audio information lies below 15 kHz in any case and little is lost by removing the top 5 kHz of the audio band. Some professional audio applications offer this rate as an option, but it is not common. It is used for the long play mode of some DAT machines, for example.

Arguments for the standardisation of higher sampling rates have become stronger in recent years, quoting evidence from sources claiming that information above 20 kHz is important for higher sound quality, or at least that the avoidance of steep filtering must be a good thing. Many sound engineers seem to be in favour of such moves, claiming to be able to distinguish clearly between the high rates and the conventional ones. One Japanese professor has shown convincing evidence that frequencies above 20 kHz stimulate the production of so-called alpha waves in the brain that correspond with a state of satisfaction and relaxation. It is certainly true that the ear's frequency response does not cut off completely at 20 kHz, but there is very limited properly supported evidence that listeners can repeatably distinguish between signals containing higher frequencies and those that do not. Whatever the difficulties of arriving at convincing evidence for all this, sufficient people believe that it matters for manufacturers to be falling over themselves to implement high rate options in their equipment and the new DVD standards incorporate such sampling frequencies as standard features. AES 5–1998 (a revision of the AES standard on sampling frequencies) now allows 96 kHz as an optional rate for applications in which the audio bandwidth exceeds 20 kHz or where relaxation of the anti-alias filtering region is desired.

Doubling the sampling frequency leads to a doubling in the overall data rate of a digital audio system and a consequent halving in storage time per megabyte. It also means that any signal processing algorithms need to process twice the amount of data and alter their algorithms accordingly, so the move to higher rates is not taken lightly in large mixing console design, for example. It follows that these higher sampling rates should be used only after careful consideration of the merits.

Low sampling frequencies such as those below 30 kHz are sometimes encountered in PC workstations for lower quality sound applications such as the storage of speech samples, the generation of internal sound effects and so forth. Multimedia applications may need to support these rates because such applications often involve the incorporation of sounds of different qualities. There are also low sampling frequency options for data reduction codecs, as discussed in Section 2.12.

2.6.3 Quantising resolution

The number of bits per sample dictates the signal-to-noise ratio or dynamic range of a digital audio system. For the time being only linear PCM systems will be considered, because the situation is different when considering systems that use non-uniform quantisation or data reduction. Table 2.3 attempts to summarise the applications for different sample resolutions.

For many years 16-bit linear PCM was considered the norm for high-quality audio applications. This is the CD standard and is capable of offering a good dynamic range of over 90 dB. For most purposes this is adequate, but it fails to reach Fielder's ideal (quoted above) of 122 dB for subjectively noise-free reproduction in professional systems. To achieve such a dynamic range requires a convertor resolution of around 21 bits, which is achievable with today's convertor technology, depending on how the specification is interpreted. Some early designs employed two convertors with a gain offset, using digital signal processing to combine the outputs of the two in the range where they overlapped, achieving a significant increase in the perceived dynamic range. Others used two convertors in parallel with independent dither, summing their outputs so that the signal rose by 6 dB but the noise by only 3 dB. So-called 24-bit convertors are indeed available today, but exactly what this means in terms of technical specification is quite hard to define. Twenty-four active bits are certainly produced at the output of such devices but their audio performance is strongly dependent

Table 2.3 Linear quantising resolution

Bits per sample	Approx. dynamic range with dither (dB)	Application
8	44	Low–moderate quality for older PC internal sound generation. Some older multimedia applications. Usually in the form of unsigned binary numbers.
12	68	Older Akai samplers, e.g. S900.
14	80	Original EIAJ format PCM adaptors, such as Sony PCM-100.
16	92	CD standard. DAT standard. Commonly used high quality resolution for consumer media, some professional recorders and multimedia PCs. Usually twos complement (signed) binary numbers.
20	116	High-quality professional audio recording and mastering applications.
24	140	Maximum resolution of most recent professional recording systems, also of AES 3 digital interface. Dynamic range exceeds psychoacoustic requirements. Hard to convert accurately at this resolution.

upon the stability of the timing clock, electrical environment, analog stages, grounding and other issues.

It is often the case that for professional recording purposes one needs a certain amount of 'headroom' – in other words some unused dynamic range above the normal peak recording level which can be used in unforeseen circumstances such as when a signal overshoots its expected level. This can be particularly necessary in live recording situations where one is never quite sure what is going to happen with recording levels. This is another reason why many professionals feel that a resolution of greater than 16 bits is desirable for original recording. For this reason, 20- and 24-bit recording formats are becoming increasingly popular, with mastering engineers then optimising the finished recording for 16-bit media (such as CD) using noise-shaped requantising processes.

At the lower quality end, older PC sound cards and internal sound generators operated at resolutions as low as 4 bits. Eight-bit resolution also used to be quite common in desktop computers, proving just about adequate for moderate quality sound through the PC's internal loudspeakers. It gave a dynamic range of nearly 50 dB undithered. Modern multimedia PCs and sound cards generally offer 16-bit resolution as standard. Some early MIDI samplers operated at 8-bit resolution, and some more recent models at 12-bit, but it is now common for MIDI samplers to offer 16- or 20-bit resolution.

2.7 Direct Stream Digital (DSD)

DSD is Sony's proprietary name for its 1-bit digital audio coding system that uses a very high sampling frequency (2.8224 MHz as a rule). This system is used for audio representation on the consumer Super Audio CD (SACD) and in various items of professional equipment used for producing SACD material. The company is trying to establish a following for this approach, for use in high-quality digital audio applications, and a number of other manufacturers are beginning to produce products that are capable of handling DSD signals. It is not directly compatible with conventional PCM systems although DSD signals can be downsampled and converted to multibit PCM if required.

DSD signals are the result of delta-sigma conversion of the analog signal, a technique used at the front end of some oversampling convertors described above. As shown in Figure 2.36, a delta-sigma convertor employs a comparator and a feedback loop containing a low pass filter that effectively quantises the difference between the current sample and the accumulated value of previous samples. If it is higher then a '1' results, if it is lower a '0' results. This creates a one-bit output that simply alternates between one and zero in a pattern that depends on the original signal waveform. Conversion to analog can be as simple a matter as passing the bit stream through a low pass filter, but is usually somewhat more sophisticated, involving noise shaping and higher order filtering.

Although one would expect one-bit signals to have an appalling signal-to-noise ratio, the exceptionally high sampling frequency spreads the noise over a very wide frequency range leading to lower noise within the audio band. Additionally, high-order noise shaping is used to reduce the noise in the audio band at the expense of that at much higher (inaudible) frequencies, as discussed earlier. A dynamic range of around 120 dB is therefore claimed, as well as a frequency response extending smoothly to over 100 kHz.

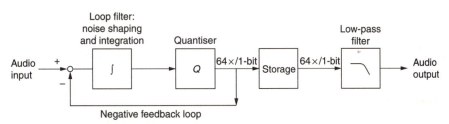

Figure 2.36 A simple example of the DSD conversion process

2.8 Changing the resolution of an audio signal (requantisation)

There may be points in an audio production when the need arises to change the resolution of a signal. A common example of this in high-quality audio is when mastering 16-bit consumer products from 20- or 24-bit recordings, but it also occurs within signal processors of all types because sample wordlengths may vary at different stages. It is important that this operation is performed correctly because incorrect requantisation results in unpleasant distortion, just like undithered quantisation in A/D conversion.

If the length of audio samples needs to be reduced then the worst possible solution is simply to remove unwanted LSBs. Taking the example of a 20-bit signal being reduced to 16 bits, one should not simply remove the 4 LSBs and expect everything to be alright. By removing the LSBs one would be creating a similar effect to not using dither in A/D conversion – in other words one would introduce low-level distortion components. Low-level signals would sound grainy and would not fade smoothly into noise. Figure 2.37 shows a 1 kHz signal at a level of −90 dBFS that originally began life at 20-bit resolution but has been truncated to 16 bits. The harmonic distortion is clearly visible.

The correct approach is to redither the signal for the target resolution by adding dither noise in the digital domain. This digital dither should be at an appropriate level for the new resolution and the LSB of the new sample should then be rounded up or down depending on the total value of the LSBs to be discarded, as shown in Figure 2.38. It is worrying to note how many low-cost digital audio applications fail to perform this operation satisfactorily, leading to complaints about sound quality. Many professional quality audio workstations allow for audio to be stored and output at a variety of resolutions and may make dither user selectable. They also allow the level of the audio signal to be changed in order that maximum use may be made of the available bits. It is normally important, for example, when mastering a CD from a 20-bit recording, to ensure that the highest level signal on the original recording is adjusted during mastering so that it peaks close to the maximum level before requantising and redithering at 16-bit resolution. In this way as much as possible of the original low-level information is preserved and quantising noise is minimised. This applies in any requantising operation, not just CD mastering. A number of applications are available that automatically scale the audio signal so that its level is optimised in this way, allowing the user to set a peak signal value up to which the highest level samples will be scaled. Since some overload detectors on digital meters and CD mastering systems look for repeated samples at maximum level to detect clipping, it is perhaps wise to set peak levels so that they

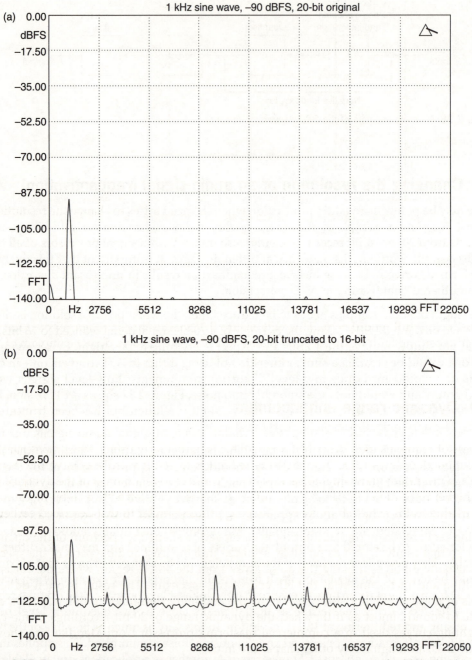

Figure 2.37 Truncation of audio samples results in distortion. (a) shows the spectrum of a 1 kHz signal generated and analysed at 20-bit resolution. In (b) the signal has been truncated to 16-bit resolution and the distortion products are clearly noticeable

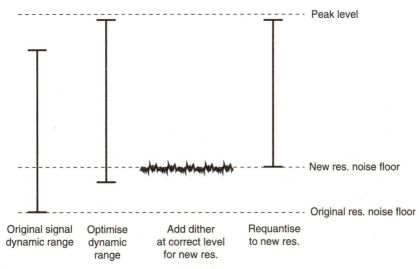

Original signal dynamic range · Optimise dynamic range · Add dither at correct level for new res. · Requantise to new res.

Figure 2.38 The correct order of events when requantising an audio signal at a lower resolution is shown here

lie just below full modulation. This will ensure that master tapes are not rejected for a suspected recording fault by duplication plants and subsequent users do not complain of 'over' levels.

2.9 Dynamic range enhancement

It is possible to maximise the subjective dynamic range of digital audio signals during the process of requantisation described above. This is particularly useful when mastering high-resolution recordings for CD because the reduction to 16-bit wordlengths would normally result in increased quantising noise. It is in fact possible to retain most of the dynamic range of a higher resolution recording, even though it is being transferred to a 16-bit medium. This remarkable feat is achieved by a noise shaping process similar to that described earlier.

During requantisation digital filtering is employed to shape the spectrum of the quantising noise so that as much of it as possible is shifted into the least audible parts of the spectrum. This usually involves moving the noise away from the 4 kHz region where the ear is most sensitive and increasing it at the HF end of the spectrum. The result is often quite high levels of noise at HF, but still lying below the audibility threshold. In this way CDs can be made to sound almost as if they had the dynamic range of 20-bit recordings. Some typical weighting curves used in a commercial mastering processor from Meridian are shown in Figure 2.39, although many other shapes are in use.

This is the principle employed in mastering systems such as Sony's Super Bit Mapping (SBM). Some approaches allow the mastering engineer to choose from a number of 'shapes' of noise until he finds one which is subjectively the most pleasing for the type of music concerned, whereas others stick to one theoretically derived 'correct' shape.

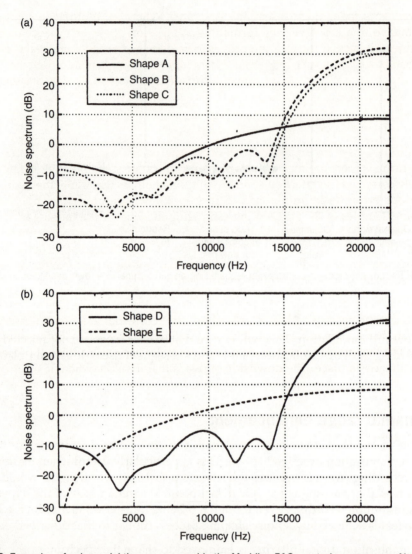

Figure 2.39 Examples of noise weighting curves used in the Meridian 518 mastering processor. Note linear frequency scale. Shape A: flat dither, 2nd order shaper. Shape B: flat dither, 9th order shaper (MAP). Shape C: flat dither, 9th order shaper (MAF). Shape D: high pass dither, 9th order shaper (MAF). Shape E: High pass dither. MAP = minimum audible pressure, MAF = minimum audible field. (Courtesy of J. R. Stuart and R. J. Wilson, Meridian Audio)

2.10 Error correction

Since this book is concerned with digital audio for workstations the topic of error correction will only be touched upon briefly. Although dedicated audio recording formats need specially designed systems to protect against the effects of data errors, systems that use computer mass storage media do not. The reason for this is that mass storage media are formatted in such a way as to make them essentially error free. When, for example, a computer disk drive is formatted at a low level, the formatting application attempts to write data to each location and read it back. If the location proves to be damaged or gives erroneous replay it is noted as a 'bad block', after which it is never used for data storage. In addition, disk and tape drives look after their own error detection and correction by a number of means that are normally transparent to the digital audio system. If a data error is detected when reading data then the block of data is normally re-read a few times to see if the data can be retrieved. The only effect of this is to slow down transfer slightly.

This differs greatly from the situation with dedicated audio formats such as DAT. In dedicated audio formats there are many levels of error protection, some of which allow errors to be completely corrected (no effect on sound quality) and others that allow the audible effects of more serious errors to be minimised. A process known as interpolation, for example, allows missing samples to be 'guessed' by estimating the level of the missing sample based on those around it (see Figure 2.40). Computer systems, on the other hand, cannot allow this type of error correction because it is assumed that data is either correct or it is useless. When reading a financial spreadsheet, for example, it would not be acceptable for an erroneous figure to be guessed by looking at those on either side!

The result is that computer mass storage media are treated as raw, error-free data storage capacity, without the need to add an overhead for error correction data once formatted. This does not mean that such media are infallible and will never give errors, because they do fail occasionally, but that audio workstations do not normally use any additional procedures on top of those already in place. The downside of this is that if an unavoidable error does arise

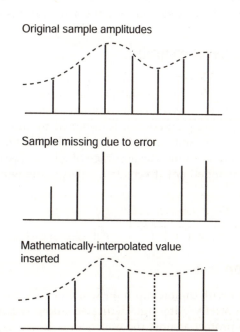

Figure 2.40 Interpolation is a means of hiding the audible effects of missing samples, as shown here

in the replay of a sound file from a digital workstation then it often results in a total inability to play that file. The file is assumed to be corrupt and the computer will not read it. The user does not have the option of being able to decide whether the error is audible, he must instead resort to one of the various computer file 'rescue packages' that attempt to rebuild the corrupted information using various proprietary techniques.

2.11 Introduction to digital audio signal processing

Just as processing operations like equalisation, fading and compression can be performed in the analog domain, so they can in the digital domain. Indeed it is often possible to achieve certain operations in the digital domain with fewer side effects such as phase distortion. It is possible to perform operations in the digital domain that are either very difficult or impossible in the analog domain. High quality, authentic-sounding artificial reverberation is one such example, in which the reflection characteristics of different halls and rooms can be accurately simulated. Digital signal processing (DSP) involves the high-speed manipulation of the binary data representing audio samples. It may involve changing the values and timing order of samples and it may involve the combining of two or more streams of audio data. DSP can affect the sound quality of digital audio in that it can add noise or distortion, although one must assume that the aim of good design is to minimise any such degradation in quality.

In the sections that follow an introduction will be given to some of the main applications of DSP in audio workstations without delving into the mathematical principles involved. In some cases the description is an over-simplification of the process, but the aim has been to illustrate concepts not to tackle the detailed design considerations involved.

2.11.1 Gain changing (level control)

It is relatively easy to change the level of an audio signal in the digital domain. It is most easy to shift its gain by 6 dB since this involves shifting the whole sample word either one step to the left or right (see Figure 2.41). Effectively the original value has been multiplied or divided by a factor of two. More precise gain control is obtained by multiplying the audio sample value by some other factor representing the increase or decrease in gain. The number of bits in the multiplication factor determines the accuracy of gain adjustment. The result of multiplying two binary numbers together is to create a new sample word which may have many

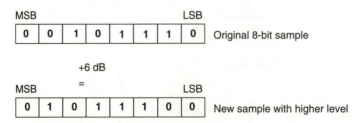

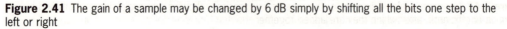

Figure 2.41 The gain of a sample may be changed by 6 dB simply by shifting all the bits one step to the left or right

more bits than the original and it is common to find that digital mixers have internal structures capable of handling 32-bit words, even though their inputs and outputs may handle only 20. Because of this, redithering is usually employed in mixers at points where the sample resolution has to be shortened, such as at any digital outputs or conversion stages, in order to preserve sound quality as described above.

The values used for multiplication in a digital gain control may be derived from any user control such as a fader, rotary knob or on-screen representation, or they may be derived from stored values in an automation system. A simple 'old-fashioned' way of deriving a digital value from an 'analog' fader is to connect the fader to a fixed voltage supply and connect the fader wiper to an A/D convertor, although it is quite common now to find controls capable of providing a direct binary output relating to their position. The 'law' of the fader (the way in which its gain is related to its physical position) can be determined by creating a suitable look-up table of values in memory which are then used as multiplication factors corresponding to each physical fader position.

2.11.2 Crossfading

Crossfading is employed widely in audio workstations at points where one section of sound is to be joined to another (edit points). It avoids the abrupt change of waveform that might otherwise result in an audible click and allows one sound to take over smoothly from the other. The process is illustrated conceptually in Figure 2.42. It involves two signals each undergoing an automated fade (binary multiplication), one downwards and the other

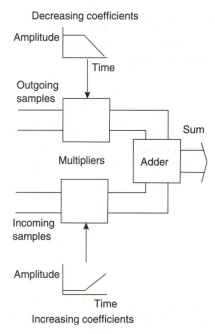

Figure 2.42 Conceptual block diagram of the crossfading process, showing two audio signals multiplied by changing coefficients, after which they are added together

upwards, followed by an addition of the two signals. By controlling the rates and coefficients involved in the fades one can create different styles of crossfade for different purposes.

2.11.3 Mixing

Mixing is the summation of independent data streams representing the different audio channels. Time coincident samples from each input channel are summed to produce a single output channel sample. Clearly it is possible to have many mix 'buses' by having a number of separate summing operations for different output channels. The result of summing a lot of signals may be to increase the overall level considerably and the architecture of the mixer must allow enough headroom for this possibility. In the same way as an analog mixer, the gain structure within a digital mixer must be such that there is an appropriate dynamic range window for the signals at each point in the chain, also allowing for operations such as equalisation that change the signal level.

2.11.4 Digital filters and equalisation

Digital filtering is something of a 'catch-all' term, and is often used to describe DSP operations that do not at first sight appear to be filtering. A digital filter is essentially a process that involves the time delay, multiplication and recombination of audio samples in all sorts of configurations, from the simplest to the most complex. Using digital filters one can create low- and high-pass filters, peaking and shelving filters, echo and reverberation effects, and even adaptive filters that adjust their characteristics to affect different parts of the signal.

To understand the basic principle of digital filters it helps to think about how one might emulate a certain analog filtering process digitally. Filter responses can be modelled in two main ways – one by looking at their frequency domain response and the other by looking at their time domain response. (There is another approach involving the so-called z-plane transform, but this is not covered here.). The frequency domain response shows how the amplitude of the filter's output varies with frequency, whereas the time domain response is usually represented in terms of an impulse response (see Figure 2.43). An impulse response shows how the filter's output responds to stimulation at the input by a single short impulse. Every

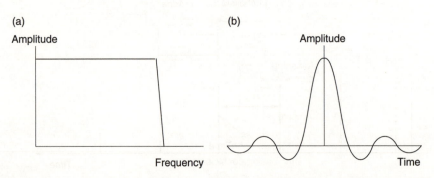

Figure 2.43 Examples of (a) the frequency response of a simple filter, and (b) the equivalent time domain impulse response

frequency response has a corresponding impulse (time) response because the two are directly related. If you change the way a filter responds in time you also change the way it responds in frequency. A mathematical process known as the Fourier transform is often used as a means of transforming a time domain response into its equivalent frequency domain response. They are simply two ways of looking at the same thing.

Digital audio is time discrete because it is sampled. Each sample represents the amplitude of the sound wave at a certain point in time. It is therefore normal to create certain filtering characteristics digitally by operating on the audio samples in the time domain. In fact if it were desired to emulate a certain analog filter characteristic digitally one would theoretically need only to measure its impulse response and model this in the digital domain. The digital version would then have the same frequency response as the analog version, and one can even envisage the possibility for favourite analog filters to be recreated for the digital workstation. The question, though, is how to create a particular impulse response characteristic digitally, and how to combine this with the audio data.

As mentioned earlier, all digital filters involve delay, multiplication and recombination of audio samples, and it is the arrangement of these elements that gives a filter its impulse response. A simple filter model is the finite impulse response (FIR) filter, or transversal filter, shown in Figure 2.44. As can be seen, this filter consists of a tapped delay line with each tap being multiplied by a certain coefficient before being summed with the outputs of the other taps. Each delay stage is normally a one sample period delay. An impulse arriving at the

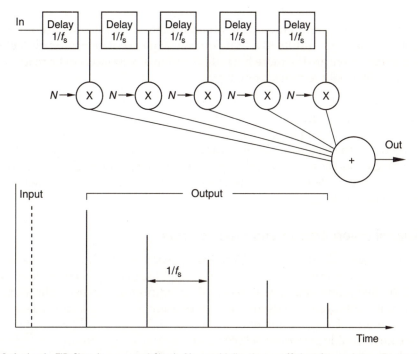

Figure 2.44 A simple FIR filter (transversal filter). N = multiplication coefficient for each tap. Response shown below indicates successive outputs samples multiplied by decreasing coefficients

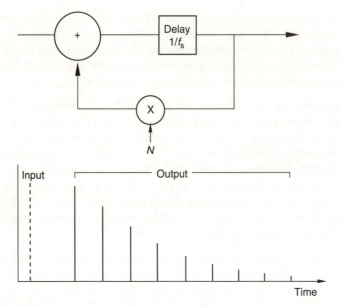

Figure 2.45 A simple IIR filter (recursive filter). The output impulses continue indefinitely but become very small. *N* in this case is about 0.8. A similar response to the previous FIR filter is achieved but with fewer stages

input would result in a number of separate versions of the impulse being summed at the output, each with a different amplitude. It is called a finite impulse response filter because a single impulse at the input results in a finite output sequence determined by the number of taps. The more taps there are the more intricate the filter's response can be made, although a simple low pass filter only requires a few taps.

The other main type is the infinite impulse response (IIR) filter, which is also known as a recursive filter because there is a degree of feedback between the output and the input (see Figure 2.45). The response of such a filter to a single impulse is an infinite output sequence, because of the feedback. IIR filters are often used in audio equipment because they involve fewer elements for most variable equalisers than equivalent FIR filters, and they are useful in effects devices. They are unfortunately not phase linear, though, whereas FIR filters can be made phase linear.

2.11.5 Digital reverberation and other effects

It can probably be seen that the IIR filter described in Section 2.11.4 forms the basis for certain digital effects, such as reverberation. The impulse response of a typical room looks something like Figure 2.46, that is an initial direct arrival of sound from the source, followed by a series of early reflections, followed by a diffuse 'tail' of densely packed reflections decaying gradually to almost nothing. Using a number of IIR filters, perhaps together with a few FIR filters, one could create a suitable pattern of delayed and attenuated versions of the original impulse to simulate the decay pattern of a room. By modifying the delays and amplitudes of the early reflections and the nature of the diffuse tail one could simulate different rooms.

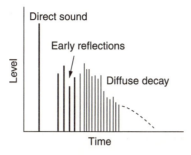

Figure 2.46 The impulse response of a typical reflective room

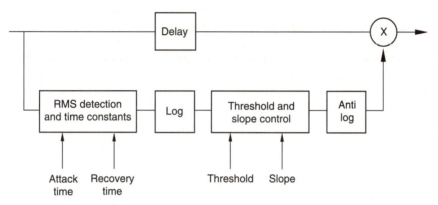

Figure 2.47 A simple digital dynamics processing operation

The design of convincing reverberation algorithms is a skilled task, and the difference between crude approaches and good ones is very noticeable. Some audio workstations offer limited reverberation effects built into the basic software package, but these often sound rather poor because of the limited DSP power available (often processed on the computer's own CPU) and the crude algorithms involved. More convincing reverberation processors are available which exist either as stand-alone devices or as optional plug-ins for the workstation, having access to more DSP capacity and tailor-made software.

Other simple effects can be introduced without much DSP capacity, such as double-tracking and phasing/flanging effects. These often only involve very simple delaying and recombination processes. Pitch shifting can also be implemented digitally, and this involves processes similar to sample rate conversion, as described below. High-quality pitch shifting requires quite considerable horsepower because of the number of calculations required.

2.11.6 Dynamics processing

Digital dynamics processing involves gain control that depends on the instantaneous level of the audio signal. A simple block diagram of such a device is shown in Figure 2.47. A side chain produces coefficients corresponding to the instantaneous gain change required, which are then used to multiply the delayed audio samples. First the r.m.s. level of the signal must be determined, after which it needs to be converted to a logarithmic value in order to determine the level change in decibels. Only samples above a certain threshold level will be affected, so a constant factor must be added to the values obtained, after which they are multiplied by a factor to represent the compression slope. The coefficient values are then antilogged to produce linear coefficients by which the audio samples can be multiplied.

2.11.7 Sample rate conversion

Sample rate conversion is necessary whenever audio is to be transferred between systems operating at different rates. The aim is to convert the audio to the new rate without any change in pitch or addition of distortion or noise. These days sample rate conversion can be a very high-quality process, although it is never an entirely transparent process because it involves modifying the sample values and timings. As with requantising algorithms, it is fairly common to encounter poorly implemented sample rate conversion on low-cost digital audio workstations, often depending very much on the specific software application rather than the hardware involved.

The easiest way to convert from one rate to another is by passing through the analog domain and resampling at the new rate, but this may introduce a small amount of extra noise. The most basic form of digital rate conversion involves the translation of samples at one fixed rate to a new fixed rate, related by a simple fractional ratio. Fractional-ratio conversion involves the mathematical calculation of samples at the new rate based on the values of samples at the old rate. Digital filtering is used to calculate the amplitudes of the new samples such that they are correct based on the impulse response of original samples, after low-pass filtering with an upper limit of the Nyquist frequency of the original sampling rate. A clock rate common to both sample rates is used to control the interpolation process. Using this method, some output samples will coincide with input samples, but only a limited number of possibilities exist for the interval between input and output samples.

If the input and output sampling rates have a variable or non-simple relationship the above does not hold true, since output samples may be required at any interval in between input samples. This requires an interpolator with many more clock phases than for fractional-ratio conversion, the intention being to pick a clock phase that most closely corresponds to the desired output sample instant at which to calculate the necessary coefficient. There will clearly be an error, which may be made smaller by increasing the number of possible interpolator phases. The audible result of the timing error is equivalent to the effects of jitter on an audio signal (see above), and should be minimised in design so that the effects of sample rate conversion are below the noise floor of the signal resolution in hand. If the input sampling rate is continuously varied (as it might be in variable-speed searching or cueing) the position of interpolated samples with relation to original samples must vary also. This requires real-time calculation of filter phase.

Many workstations now include sample rate conversion as either a standard or optional feature, so that audio material recorded and edited at one rate can be reproduced at another. It is important to ensure that the quality of the sample rate conversion is high enough not to affect the sound quality of your recordings, and it should only be used if it cannot be avoided.

Poorly implemented applications sometimes omit to use correct low-pass filtering to avoid aliasing, or incorporate very basic digital filters, resulting in poor sound quality after rate conversion.

Sample rate conversion is also useful as a means of synchronising an external digital source to a standard sampling frequency reference, when it is outside the range receivable by a workstation.

2.12 Audio data reduction

Conventional PCM audio has a high data rate, and there are many applications for which it would be an advantage to have a lower data rate without much (or any) loss of sound quality. Sixteen-bit linear PCM at a sampling rate of 44.1 kHz ('CD quality digital audio') results in a data rate of about 700 kbit s^{-1}. For multimedia applications, broadcasting, communications and some consumer purposes (e.g. streaming over the Internet) the data rate may be reduced to a fraction of this with minimal effect on the perceived sound quality. At very low rates the effect on sound quality is traded off with the bit rate required. Simple techniques for reducing the data rate, such as reducing the sampling rate or number of bits per sample, would have a very noticeable effect on sound quality, so most modern low bit rate coding works by exploiting the phenomenon of auditory masking to 'hide' the increased noise resulting from bit rate reduction in parts of the audio spectrum where it will hopefully be inaudible. There are a number of types of low bit rate coding used in audio systems, working on similar principles, and used for applications such as consumer disk and tape systems (e.g. Sony ATRAC), digital cinema sound (e.g. Dolby Digital, Sony SDDS, DTS) and multimedia applications (e.g. MPEG).

2.12.1 Why reduce the data rate?

Nothing is inherently wrong with linear PCM from a sound quality point of view, indeed it is probably the best thing to use. The problem is simply that the data rate is too high for a number of applications. Two channels of linear PCM require a rate of around 1.4 Mbit s^{-1}, whereas applications such as digital audio broadcasting (DAB) or digital radio need it to be more like 128 kbit s^{-1} (or perhaps lower for some applications) in order to fit sufficient channels into the radio frequency spectrum – in other words more than ten times less data per second. Some Internet streaming applications need it to be even lower than this, with rates down in the low tens of kilobits per second for modem-oriented connections or mobile communications.

The efficiency of mass storage media and data networks is related to their data transfer rates. The more data can be moved per second, the more audio channels may be handled simultaneously, the faster a disk can be copied, the faster a sound file can be transmitted across the world. In reducing the data rate that each audio channel demands, one also reduces the requirement for such high specifications from storage media and networks, or alternatively one can obtain greater functionality from the same specification. A network connection capable of handling eight channels of linear PCM simultaneously could be made to handle, say, 48 channels of data-reduced audio, without unduly affecting sound quality.

Although this sounds like magic and makes it seem as if there is no point in continuing to use linear PCM, it must be appreciated that the data reduction is achieved by throwing away data from the original audio signal. The more data is thrown away the more likely it is that unwanted audible effects will be noticed. The design aim of most of these systems is to try to retain as much as possible of the sound quality whilst throwing away as much data as possible, so it follows that one should always use the least data reduction necessary, where there is a choice.

2.12.2 Lossless and lossy coding

There is an important distinction to be made between the type of data reduction used in some computer applications and the approach used in many audio coders. The distinction is really between 'lossless' coding and coding which involves some loss of information (see Figure 2.48). It is quite common to use data compression on computer files in order to fit more information onto a given disk or tape, but such compression is usually lossless in that the original data are reconstructed bit for bit when the file is decompressed. A number of tape backup devices for computers have a compression facility for increasing the apparent capacity of the medium, for example. Methods are used which exploit redundancy in the information, such as coding a string of eighty zeros by replacing them with a short message stating the value of the following data and the number of bytes involved. This is particularly relevant in single-frame bit-mapped picture files where there may be considerable runs of black or white in each line of a scan, where nothing in the image is changing. One may expect files compressed using off-the-shelf PC data compression applications to be reduced to perhaps 25–50 per cent of their original size, but it must be remembered that they are often dealing with static data, and do not have to work in real time. Also, it is not normally acceptable for decompressed computer data to be anything but the original data.

It is possible to use lossless coding on audio signals. Lossless coding allows the original PCM data to be reconstructed perfectly by the decoder and is therefore 'noiseless' since there is no effect on audio quality. The data reduction obtained using these methods ranges from nothing to about 2.5:1 and is variable depending on the program material. This is because

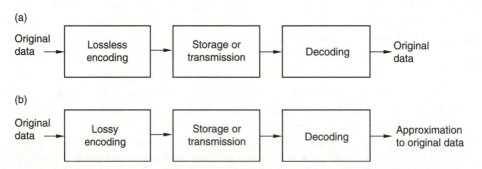

Figure 2.48 (a) In lossless coding the original data is reconstructed perfectly upon decoding, resulting in no loss of information. (b) In lossy coding the decoded information is not the same as that originally coded, but the coder is designed so that the effects of the process are minimal

audio signals have an unpredictable content, do not make use of a standard limited character set, and do not spend long periods of time in one binary state or the other. Although it is possible to perform this reduction in real time, the coding gains are not sufficient for many applications. Nonetheless, a halving in the average audio data rate is certainly a useful saving. A form of lossless data reduction known as Direct Stream Transfer (DST) can be used for Super Audio CD (see Section 2.7) in order to fit the required multichannel audio data into the space available. A similar system is available for DVD-Audio, called MLP (Meridian Lossless Packing), discussed further in Chapter 8.

'Noisy' or lossy coding methods make possible a far greater degree of data reduction, but require the designer and user to arrive at a compromise between the degree of data reduction and potential effects on sound quality. Here data reduction is achieved by coding the signal less accurately than in the original PCM format (using fewer bits per sample), thereby increasing quantising noise, but with the intention that increases in noise will be 'masked' (made inaudible) by the signal. The original data is not reconstructed perfectly on decoding. The success of such techniques therefore relies on being able to model the characteristics of the human hearing process in order to predict the masking effect of the signal at any point in time – hence the common term 'perceptual coding' for this approach. Using detailed psychoacoustic models it is possible to code high-quality audio at rates under 100 kbit s^{-1} per channel with minimal effects on audio quality. Higher data rates, such as 192 kbit s^{-1}, can be used to obtain an audio quality that is demonstrably indistinguishable from the original PCM.

2.12.3 MPEG – an example of lossy coding

The following is a very brief overview of how one approach works, based on the technology involved in the MPEG (Moving Pictures Expert Group) standards.

As shown in Figure 2.49, the incoming digital audio signal is filtered into a number of narrow frequency bands. Parallel to this a computer model of the human hearing process (an auditory model) analyses a short portion of the audio signal (a few milliseconds). This analysis is used to determine what parts of the audio spectrum will be masked, and to what degree, during that short time period. In bands where there is a strong signal, quantising noise can be allowed to rise considerably without it being heard, because one signal is very efficient at masking another lower level signal in the same band as itself (see Figure 2.50). Provided that the noise is kept below the masking threshold in each band it should be inaudible.

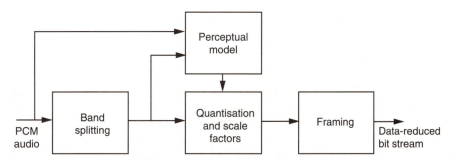

Figure 2.49 Generalised block diagram of a psychoacoustic low bit rate coder

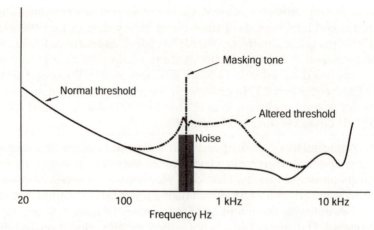

Figure 2.50 Quantising noise lying under the masking threshold will normally be inaudible

Blocks of audio samples in each narrow band are scaled (low level signals are amplified so that they use more of the most significant bits of the range) and the scaled samples are then reduced in resolution (requantised) by reducing the number of bits available to represent each sample – a process that results in increased quantising noise. The output of the auditory model is used to control the requantising process so that the sound quality remains as high as possible for a given bit rate. The greatest number of bits is allocated to frequency bands where noise would be most audible, and the fewest to those bands where the noise would be effectively masked by the signal. Control information is sent along with the blocks of bit-rate-reduced samples to allow them to be reconstructed at the correct level and resolution upon decoding.

The above process is repeated every few milliseconds, so that the masking model is constantly being updated to take account of changes in the audio signal. Carefully implemented, such a process can result in a reduction of the data rate to anything from about one-quarter to less than one-tenth of the original data rate. A decoder uses the control information transmitted with the bit-rate-reduced samples to restore the samples to their correct level and can determine how many bits were allocated to each frequency band by the encoder, reconstructing linear PCM samples and then recombining the frequency bands to form a single output (see Figure 2.51). A decoder can be much less complex, and therefore cheaper, than an encoder, because it does not need to contain the auditory model.

A standard known as MPEG-1, published by the International Standards Organisation (ISO 11172–3), defines a number of 'layers' of complexity for low bit rate audio coders as shown in Table 2.4. Each of the layers can be operated at any of the bit rates within the ranges shown (although some of the higher rates are intended for stereo modes) and the user must make appropriate decisions about what sound quality is appropriate for each application. The lower the data rate, the lower the sound quality that will be obtained. At high data rates the encoding-decoding process has been judged by many to be audibly 'transparent' – in other words listeners cannot detect that the coded and decoded signal is different from the original input. The target bit rates were for 'transparent' coding.

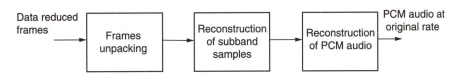

Figure 2.51 Generalised block diagram of an MPEG-Audio decoder

Table 2.4 MPEG-1 layers

Layer	Complexity	Min. delay (ms)	Bit rate range (kbit s^{-1})	Target (kbit s^{-1})
1	Low	19	32–448	192
2	Moderate	35	32–384*	128
3	High	59	32–320	64

* In Layer 2, bit rates of 224 kbit s^{-1} and above are for stereo modes only.

'MP3' will be for many people the name associated with downloading music files from the Internet. The term MP3 has caused some confusion; it is short for MPEG-1 Layer 3, but MP3 has virtually become a generic term for the system used for receiving compressed audio from the Internet. There is also MPEG-2 which can handle multichannel surround, and further developments in this and later systems will be briefly touched upon.

MPEG-2 BC (Backwards Compatible with MPEG-1) additionally supports sampling frequencies from 16 kHz to 22.05 kHz and 24 kHz at bit rates from 32 to 256 kbit s^{-1} for Layer 1. For Layers 2 and 3, bit rates are from 8 to 160 kbit s^{-1}. Developments intended to supersede MPEG-2 BC have included MPEG-2 AAC (Advanced Audio Coding). This defines a standard for multichannel coding of up to 48 channels, with sampling rates from 8 kHz to 96 kHz. It also incorporates a Modified Discrete Cosine transform system as used in the MiniDisc coding format (ATRAC). MPEG-2 AAC was not, however, designed to be backwards compatible with MPEG-1.

MPEG-4 'natural audio coding' is based on the standards outlined for MPEG-2 AAC; it includes further coding techniques for reducing transmission bandwidth and it can scale the bit rate according to the complexity of the decoder. There are also intermediate levels of parametric representation in MPEG-4 such as used in speech coding, whereby speed and pitch of basic signals can be altered over time. One has access to a variety of methods of representing sound at different levels of abstraction and complexity, all the way from natural audio coding (lowest level of abstraction), through parametric coding systems based on speech synthesis and low level parameter modification, to fully synthetic audio objects.

When audio signals are described in the form of 'objects' and 'scenes', it requires that they be rendered or synthesised by a suitable decoder. Structured Audio (SA) in MPEG-4 enables synthetic sound sources to be represented and controlled at very low bit rates (less than 1 kbit s^{-1}). An SA decoder can synthesise music and sound effects. SAOL (Structured Audio Orchestra Language), as used in MPEG-4, was developed at MIT and is an evolution of CSound (a synthesis language used widely in the electroacoustic music and academic communities). It enables 'instruments' and 'scores' to be downloaded. The instruments define the parameters of a number of sound sources that are to be rendered by synthesis (e.g. FM, wavetable, granular, additive) and the 'score' is a list of control information that governs what those instruments play and when (represented in the SASL or Structured Audio Score Language format). This is rather like a more refined version of the established MIDI control protocol, and indeed MIDI can be used if required for basic music performance control. This is discussed further in Chapter 4.

Sound scenes, as distinct from sound objects, are usually made up two elements – that is the sound objects and the environment within which they are located. Both elements are integrated within one part of MPEG-4. This part of MPEG-4 uses so-called BIFS (Binary Format for Scenes) for describing the composition of scenes (both visual and audio). The objects are known as nodes and are based on VRML (virtual reality modelling language). So-called Audio BIFS can be post-processed and represent parametric descriptions of sound objects. Advanced Audio BIFS also enable virtual environments to be described in the form of perceptual room acoustics parameters, including positioning and directivity of sound objects. MPEG-4 audio scene description distinguishes between physical and perceptual representation of scenes, rather like the low- and high-level description information mentioned above.

2.12.4 Other data-reduced formats

Dolby Digital or AC-3 encoding was developed as a means of delivering 5.1-channel surround to cinemas or the home without the need for analog matrix encoding. The AC-3 coding algorithm can be used for a wide range of different audio signal configurations and bit rates from 32 kbit s^{-1} for a single mono channel up to 640 kbit s^{-1} for surround signals. It is used widely for the distribution of digital sound tracks on 35 mm movie films, the data being stored optically in the space between the sprocket holes on the film.

It is sufficient to say here that the process involves a number of techniques by which the data representing audio from the source channels is transformed into the frequency domain and requantised to a lower resolution, relying on the masking characteristics of the human hearing process to hide the increased quantising noise that results from this process. A common bit pool is used so that channels requiring higher data rates than others can trade their bit rate requirements provided that the overall total bit rate does not exceed the constant rate specified.

Aside from the representation of surround sound in a compact digital form, Dolby Digital includes a variety of operational features that enhance system flexibility and help adapt replay to a variety of consumer situations. These include dialogue normalisation ('dialnorm') and the option to include dynamic range control information alongside the audio data for use in environments where background noise prevents the full dynamic range of the source material to be heard. Downmix control information can also be carried alongside the audio data in order that a two-channel version of the surround sound material can be reconstructed in the decoder. As a rule, Dolby Digital data is stored or transmitted with the highest number of channels needed for the end product to be represented and any compatible downmixes are created in the decoder. This differs from some other systems where a two-channel downmix is carried alongside the surround information.

The DTS (Digital Theater Systems) 'Coherent Acoustics' system is another digital signal coding format that can be used to deliver surround sound in consumer or professional applications, using low bit rate coding techniques to reduce the data rate of the audio information. The DTS system can accommodate a wide range of bit rates from 32 kbit s^{-1} up to 4.096 Mbit s^{-1} (somewhat higher than Dolby Digital), with up to eight source channels and with sampling rates up to 192 kHz. Variable bit rate and lossless coding are also optional. Downmixing and dynamic range control options are provided in the system. Because the maximum data rate is typically somewhat higher than that of Dolby Digital or MPEG, a greater margin can be engineered between the signal and any artefacts of low bit rate coding, leading to potentially higher sound quality. Such judgements, though, are obviously up to the individual and it is impossible to make blanket statements about comparative sound quality between systems.

SDDS stands for Sony Dynamic Digital Sound, and is the third of the main competing formats for digital film sound. Using Sony's ATRAC data reduction system (also used on MiniDiscs), it too encodes audio data with a substantial saving in bit rate compared with the original PCM (about 5:1 compression).

Real Networks has been developing data reduction for Internet streaming applications for a number of years and specialises in squeezing the maximum quality possible out of very low bit rates. It has recently released 'Real Audio with ATRAC 3' which succeeds the earlier Real Audio G2 standard. Audio can be coded at rates between 12 and 352 kbit s^{-1}, occupying only 63 per cent of the bandwidth previously consumed by G2.

Further reading

Bosi, M. and Goldberg, R. (2003) *Introduction to Digital Audio Coding and Standards*. Kluwer Academic Publishers.

Watkinson, J. (2001) *The Art of Digital Audio,* third edition. Focal Press.

3 Recording, replay and editing principles

This chapter is concerned with the principles of audio recording and replay using mass storage media, including various approaches to editing.

3.1 The sound file

In audio workstations, recordings are stored in sound files on mass storage media. The storage medium is normally a disk but other media may be used in certain circumstances such as for backup of disks. For the sake of simplicity disks are assumed to be the primary means of storage in the following sections. A sound file is an individual recording of any length from a few seconds to a number of hours (within the limits of the system). With tape recording, parts of a tape may be recorded at different times, and in such a situation there will be sections of that tape that represent distinctly separate recordings: they may be 'tracks' for an album, 'takes' of a recording session, or short individual sounds such as sound effects. This is the closest that tape recording gets to the concept of the sound file: that is a distinct unit of recorded audio, the size of the unit being anything that fits into the available space.

In the audio workstation the disk can be thought of as a 'sound store' in which no one part has any specific time relationship to any other part – no section can be said to be 'before' another or 'after' another. This is the nature of random- or direct-access storage (although some forms of optical disk store data contiguously for all or part of their capacity, although they retain random accessibility). It has led to the use of the somewhat confusing description 'non-linear recording', which contrasts with the 'linear' recording process that takes place on tape. (To many people, the term 'non-linear' means that the audio has been quantised non-linearly, which is not the case in most professional audio systems.)

A disk may accommodate a number of sound files of different lengths. It is possible that one file might be a 10 minute music track whilst another might be a 1 second sound effect. As many sound files can be kept in the store as will fit in the space available, although some operating systems have upper limits on the number of individual files that can be handled by the directory structure. Each sound file is made up of a number of discrete data blocks and normally the block size will limit the minimum size occupied by a file since systems do not normally write partial blocks (see below).

Normally sound files are either mono or stereo – that is either a single channel or two related channels of audio combined into one file. They are rarely more than stereo, since multichannel operation is normally achieved by storing a number of separate mono files, one for each channel. Stereo sound files contain the left and right channels of a stereo pair, usually interleaved on a sample-by-sample basis as described in Chapter 6, and are useful when the two channels will always be replayed together and in a fixed timing relationship. Accessing a stereo file is then no different from accessing a mono file, except that the stereo file requires twice the amount of data to be transferred for the same duration of audio. As far as the user is concerned, the system can present a stereo sound file under a single title and note in the file header that it is stereo. In this case any buffering (see below) would have to be split such that left channel samples would be written to and read from one group of memory addresses, and right channel samples to and from another. As would be expected, stereo files take up twice the amount of disk space of the equivalent mono file.

3.2 RAM buffering

Computer disk drives were not originally designed for recording audio, although they can be made to serve this purpose. As explained in Chapter 5, a disk is normally formatted in sectors, often grouped into blocks, and the blocks making up a file need not be stored contiguously (contiguous means physically adjacent). The result of this is that data transfer to and from such media is not smooth but intermittent or burst-like. Furthermore, editing may involve the joining of sections from files stored in physically separate locations, resulting in breaks in the data flow from disk at edit points whilst the new file is located. Although this burst transfer rarely presents a problem in applications such as text processing (it does not matter if a text file is loaded in bursts) it is unsuitable for the recording and replay of real-time audio. Audio (and video in most cases) requires that samples are transferred to and from convertors or digital interfaces at a constant rate, in an unbroken stream. Consequently digital audio hardware and software must include mechanisms for converting burst data flow into continuous data flow and vice versa. This is achieved by using RAM (random access memory) as a short-term 'buffer' or reservoir.

RAM is temporary solid-state memory with a very fast access time and transfer rate. It can be addressed directly by the processing hardware of the audio workstation, and is used as an intermediate store for audio samples on their way to and from the disk drive (see Figure 3.1). During recording, audio samples are written into the RAM at a regular rate and read out again a short time later to be written as blocks of data on the disk. At least one complete sector of audio is transferred in one operation, and usually a number of sectors are written in one operation (see Section 3.4). The transfer is effectively time compressed, since samples acquired over, say, 100 ms, may be written to the disk in a short burst lasting only 20 ms, followed by a gap. During simple replay, data blocks are transferred from the disk into RAM

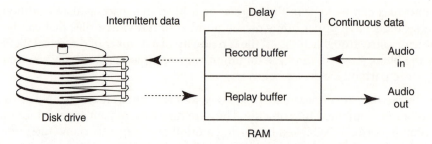

Figure 3.1 RAM buffering is used to convert burst data flow to continuous data flow, and vice versa

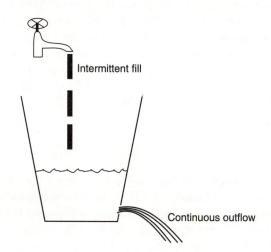

Figure 3.2 RAM buffering may be likened to a water reservoir that acts to convert intermittent filling to continuous outflow

in bursts and then read out at a steady rate for transfer to a D/A convertor or digital interface. The process of transferring out from the buffer normally begins before the file has been transferred completely into the buffer, because otherwise (a) there would be an unacceptable delay between the initiation of replay and the onset of an audible output, and (b) the size of the buffer would have to be great enough to hold the largest sound file entirely.

The RAM buffer acts in a similar way to a water reservoir. It allows supply and demand to vary at its input and its output whilst remaining able to provide an unbroken supply, assuming that sufficient water remains in the reservoir. Figure 3.2 shows an analogy with a water bucket that has a hole in the bottom, filled by a tap. One may liken the tap to a disk drive and the water flowing out of the hole to an audio output. The tap may fill the bucket in bursts, but within certain limits this is converted into continuous outflow. Provided that the average flow rate of water entering the bucket is the same as the average rate at which it flows out of the hole, then the bucket will neither empty nor overflow (within the limits of the size of the bucket). If water flows out of the hole faster than it is supplied by the tap then the bucket will

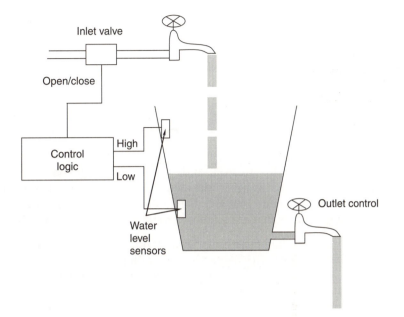

Figure 3.3 A control system could be added to the simple reservoir to regulate inflow and outflow so that supply and demand are linked

eventually become empty. On the other hand, the bucket could overflow if the tap was left on all the time and was filling the bucket faster than the hole could empty it.

Clearly some control mechanism is called for. Sensors could be attached to the insides of the bucket to detect high and low water levels, as shown in Figure 3.3, connected to control logic which operated a valve in the supply line. The valve would be opened when the water level was getting low, and closed when it was getting high. A tap on the bucket outlet could be added to stop and start the flow (the equivalent of the PLAY button for audio replay). Equivalents of this control mechanism exist in audio workstation software. Pointers are incremented up and down to register the state of fullness of RAM buffers during record and replay operations, and action is taken at certain states of fullness either to transfer new blocks of data to and from the disk or to halt transfer.

The analogy can be taken further. There might be more than one hole in the bucket (more audio outputs), larger holes in the bucket (higher sampling rates and resolutions) or a tap with low water pressure (a slow storage device). Audio system design is largely a matter of juggling with these parameters and others to optimise the system performance. (The bucket analogy does not hold water if examined too closely, as water will flow faster out of the holes in the bucket the fuller the bucket, and this does not hold true for memory buffers in audio workstations!)

RAM buffering has a number of other uses. Firstly, it can be used to ensure that any short term timing irregularities in the data coming from the storage device will be ironed out and will not be allowed to affect audio quality. Data written into memory from the store, even if it has timing jitter, can be read out from the store at a constant steady rate, under control of an accurate crystal clock. The only penalty of buffering is that it introduces a small delay between the input to and the output from the buffer, the extent of which depends on the delay between the writing of samples to the RAM and the reading of them out again. The maximum delay is limited by the size of the buffer, as with a small buffer there will come a point where the memory is filled and must be partially emptied before any new samples can be written in. The delay effect of the buffer can be disguised in operation because data can be read from disk ahead of the required time and written at an appropriate time after sample acquisition.

Secondly, the buffer may be used for synchronisation purposes. If audio data is to be synchronised with an external reference such as timecode, then the rate at which data is read out of the buffer can be finely adjusted to ensure that lock is maintained. It is also possible to align the timings of multiple audio channels that are supposed to be nominally in sync with each other.

The size of buffer in a digital audio system may or may not be under the user's control, but is typically in the region of 0.5–2 Mbytes. An area of operating RAM will be set aside for this purpose, sometimes located on the audio processing board itself rather than being system RAM of the host computer. Generally, the more channels to be handled, the larger the buffer, since each channel requires its own memory space; also a larger buffer can help to compensate for badly fragmented storage space (see Section 5.4), although it cannot make up for a disk drive that is too slow overall.

3.3 Disk drive performance issues

Access time and transfer rate are important features governing the suitability of disk drives for primary digital audio storage. The sustained transfer rate is far more important than the instantaneous rate, since this is more likely to represent the performance in real file transfer operations.

Tables 3.1 and 3.2 show the data rates and capacities required for different resolutions of digital audio, either linear PCM or data-reduced. From this one can begin to work out the

Table 3.1 Data rates and capacities for linear PCM

Sampling rate (kHz)	Resolution (bits)	Bit rate (kbit s^{-1})	Capacity/min. (Mbytes min^{-1})	Capacity/hour (Mbytes $hour^{-1}$)
96	16	1536	11.0	659
88.1	16	1410	10.1	605
48	20	960	6.9	412
48	16	768	5.5	330
44.1	16	706	5.0	303
44.1	8	353	2.5	151
32	16	512	3.7	220
22.05	8	176	1.3	76
11	8	88	0.6	38

Table 3.2 Data rates and capacities for data-reduced audio

Bit rate (kbit s^{-1})	Capacity/min. (Mbytes min^{-1})	Capacity/hour (Mbytes hour^{-1})
64	0.5	27
96	0.7	41
128	0.9	55
196	1.4	84
256	1.8	110
384	2.7	165

performance requirements of storage devices. The data rate for one channel of audio at 48 kHz, 16 bits, amounts to around 0.75 Mbit s^{-1}, thus it might be assumed that a device with a transfer rate of 0.75 Mbit s^{-1} would be able to handle the replay of one audio channel's data satisfactorily. If the store were made up of solid state RAM which has a negligible access time (of the order of tens or hundreds of nano seconds) then a transfer rate of 0.75 Mbit s^{-1} would be adequate, but in the usual case where the store is a disk drive, the access time will severely limit the average transfer rate. Although the burst transfer rate from the disk to the buffer may be high, the gaps between transfers as the drive searches for new blocks of data will reduce the effective rate. It is therefore the combination of access time and transfer rate that go to make up the effective transfer rate. What is needed is a fast transfer rate and a fast access time.

The job of the buffer is to disguise the effects of access time delays, and it may be seen that the size of the buffer will depend on the potential access delay, among other things. If transfer is erratic, that is with long gaps and then extremely fast transfers, the buffer is likely to swing between being very full and very empty, rather than deviating a small amount around a half full position. In the former case it is likely that a larger buffer will be required.

Over a period of time the disk is likely to become fragmented and this will lead to file blocks being stored in a number of physically separate locations. The more fragmented a store becomes the lower the efficiency of data retrieval, as a file will be transferred in a number of short bursts separated by breaks while the next block is accessed. Furthermore, the access time depends on how far apart the blocks are, as the retrieval mechanism will take less time to travel a short distance than to travel a long way. (This is covered further in Chapter 5.) For this reason, figures quoted for access time can only ever be a rough guide.

Certain storage media have different access times and transfer rates when recording (writing) to those encountered when replaying (reading). For example, hard disks use a magnetic recording method that overwrites old information completely without erasing it first. Some magneto-optical drives require a two-stage process in order to rewrite over old data, so the required block must be erased on one revolution and then written on the next. There may also be a 'verify' pass after writing. This suggests that recording performance may not always be as good as replay performance, and that a disk drive may be able to replay more channels simultaneously than it can record.

For all the above reasons it is often difficult to calculate how many channels of audio one may expect a disk drive to be able to handle. To take an example, assume an older disk drive with an average access time of 20 milliseconds and a transfer rate of 20 Mbit s^{-1}. If the access time was near zero then the transfer rate of 20 Mbit s^{-1} would allow around 26 channels of audio to be transferred at the example resolution given above, but the effective transfer rate in real operation will bring this number down to perhaps twelve or fewer channels for safe, reliable operation in a wide variety of operational circumstances. Editing operations also place considerable additional demands on disk drive performance, depending on how edits are carried out. Because of all this, some manufacturers play very safe and limit their systems to a small number of channels per disk drive, even if the drive might be able to handle more under some circumstances. Other software simply leaves it up to the user to determine when disk drive will fail to perform, or provides a warning when it is getting close to the limit. The effect of using a disk drive beyond the limits of its performance is normally to experience 'drop-outs' in replay, and system messages such as 'drive too slow' when attempting to replay large numbers of channels with many edits.

3.4 Allocation units or transfer blocks

Optimising the efficiency of data transfer to and from a storage device will depend on keeping the number of head seeks to a minimum for any given file transfer. This requires careful optimisation of the size and position of the audio transfer blocks or allocation units. Typically, a disk sector (that is the smallest addressable storage unit) contains 512 bytes of information, although some drives use 1024 byte (or greater) sectors. This is very small in relation to the size of a digital audio file of even moderate length and if a file were to be split up into chunks of 512 bytes spread all over the disk then efficiency would be impossibly reduced due to the large number of seeks required to different parts of the disk. For this reason a minimum transfer block or allocation unit is usually defined, which is a certain number of bytes that are transferred together and preferably stored contiguously in order to improve efficiency. It might be that a transfer block would contain 8 kbytes of audio data, which in the case of 512 byte sectors would correspond to 16 sectors. The size of the transfer block must be small enough to engender efficient use of the disk space in cases of fragmentation and large enough to result in efficient data transfer. If the digital audio system stores audio under the native filing system of the host computer then the size of the transfer block may be fixed during the formatting of the disk volume. A common size is 32 kbytes.

3.5 Multichannel recording and replay

3.5.1 Multitrack or multichannel?

It is important to understand a fundamental difference between the workstation concept of multichannel operation and the traditional concept of multitrack tape recording. The difference is that 'tracks' and 'channels' need not necessarily mean the same thing. In a multitrack tape recorder there may be up to 48 tracks of audio recorded onto the tape, each of which is an independent mono track lasting the length of the tape. Each of these tracks feeds a numbered audio output and is fed from a numbered input. Once sound is recorded onto a numbered

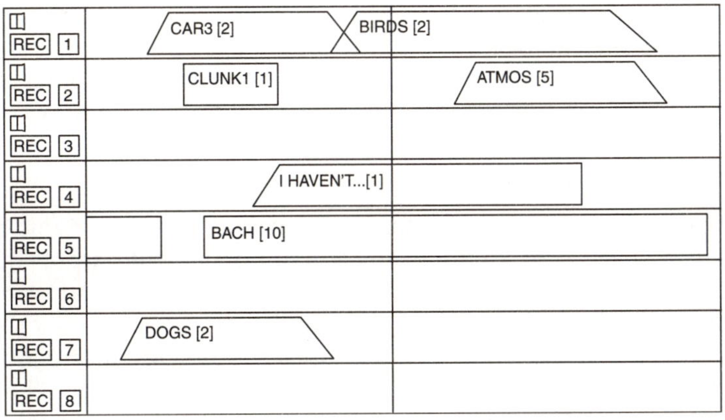

Figure 3.4 Tracks are represented in this simulated display as horizontal bands containing named sound file segments. The output to which that track is routed is selected at the left-hand side, along with recording and replay muting controls

track it is fixed in time and physical position in relation to other sounds recorded on the same tape and it will be replayed on the same-numbered audio channel at all times (unless the internal wiring of the machine is changed).

In workstations the terms 'track' and 'channel' may be separated from each other, in that a sound file, once stored, may be replayed on any audio channel depending on the user's choice. It may even be that the concept of the track is done away with altogether, but this depends on the user interface of the system. Most manufacturers have chosen to retain the concept of tracks because it is convenient and well understood. Tracks, in workstation terminology, are just ways of showing which sound elements have been grouped together for replay on the same channel, but are not fixed as in tape recording. Figure 3.4 shows a simulated display from a multitrack package in which tracks are represented as horizontal bands containing sound file segments. On the left-hand side it is possible to change the physical audio output assigned for replay of that track. The sound segments can be moved around in time on the virtual track by sliding them left or right and they can be copied or moved to other tracks if necessary.

3.5.2 Inputs, outputs, tracks and channels

Because of the looser relationship between tracks, channels and audio inputs and outputs, confusion occasionally arises. Firstly, none of these are necessarily related to each other, although a designer may decide to relate them. In a 24-track tape machine, there are 24 inputs, 24 outputs, 24 tracks and 24 channels, so it is very easy to see a direct relationship between one and the others. It is even possible to say exactly where on the tape track 13 will be recorded at any point in time. In a workstation it is possible, for example, for there to be two inputs, eight outputs, 99 tracks, and eight channels. It is rarely possible to say exactly where track 13 will be recorded at any point, or what information is recorded on it, as it all depends on what the user has decided. In this example it may be that only two inputs have been provided because that is all that the designer is going to allow you to record at any one time, but it is highly likely that these two inputs could be routed to any 'track' or any output channel. The two inputs allow for the recording of stereo or mono sound files that will be stored in a free location and given names by the user. Although only two 'tracks' may be recorded at once, this operation may be performed many times to build up a large number of sound files in the store.

In some systems, the concept of the track has been considered as important, and in the above example there are 99 tracks (just a virtual concept) but only eight outputs or channels. This is because the user is allowed to record information onto any of the tracks, but he may only replay eight of them simultaneously. The number of simultaneous output channels is limited by the transfer rates of the storage devices, the signal processing capacity of the system and the number of D/A convertors or digital outputs employed. By expanding the system, adding more or faster disks and adding more processing power, more of the 99 tracks could be replayed simultaneously. Many manufacturers have taken this modular approach to system design, allowing the user to start off in a small way, expanding the capabilities of the system as time and money allow.

3.5.3 Track usage, storage capacity and disk assignment

The storage space required for multiple channels increases pro rata with the number of channels, although in fact eight-track recording may not require eight times the storage space of mono recording because many 'tracks' may be blank for large amounts of the time. If you think about the average multitrack recording on tape you will realise that many tracks have large gaps with nothing recorded. The total storage space used will depend on the total duration of the mono sound files used in the program, whatever tracks or channels they are assigned to. It has been estimated, for example, that sound effects tracks in feature film production contain about two-thirds silence and that dialogue tracks are only 10–20 per cent utilised.

(It is often said that disk-based systems do not record the silences on tracks and therefore do not use up as much storage space as might be expected, but the only time when silences save storage time is when they exist as blank spaces between the output of sound files, where no sound file is assigned to play (see Figure 3.5). Recorded silence uses as much disk space as recorded music!)

Multichannel disk recording systems sometimes use more than one disk drive, and there is a limit to the number of channels that can be serviced by a single drive. It is necessary, therefore, to determine firstly how many channels a storage device will handle realistically and then to work out how many are needed to give the total capacity required. Some older systems attempted to imitate a multitrack tape recorder in assigning certain disk drives permanently to certain groups of tracks, as shown in Figure 3.6, but this limited operational flexibility. If a sound file from one track were needed on another it might have to be copied to the appropriate drive, which would take time. This approach is becoming much less common now that the performance of disk drives is getting to the point where one can replay perhaps 16 channels simultaneously from a single drive. In modular systems, if one needs greater channel recording and replay capacity one can simply add further disk I/O cards, each connected to a

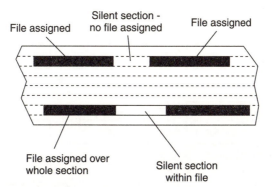

File assigned   Silent section -   File assigned
no file assigned

File assigned over   Silent section
whole section   within file

Figure 3.5 The silent section on the upper track does not require any disk space because no recording exists for this time slot. The silent section on the lower track was recorded as part of a file, and so consumes as much space as any other sound

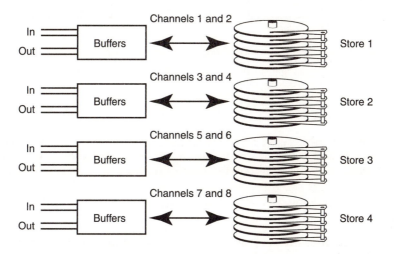

Figure 3.6 Some older multitrack disk systems assigned disks permanently to certain tracks, as shown here

separate disk. If one needs more storage capacity then more disk drives can be attached to the same SCSI bus, as shown in Figure 3.7. It is then relatively unimportant which drive a file is stored on, provided that the software is capable of handling the addressing of multiple drives. There may be some restrictions if the user has constructed a play list which requires more simultaneous file transfers from a certain disk than can be handled.

3.5.4 Dropping-in

In multitrack music systems the capability to 'drop-in' is important. Dropping-in involves instantaneous entry into record mode at the touch of a button and it is expected that a

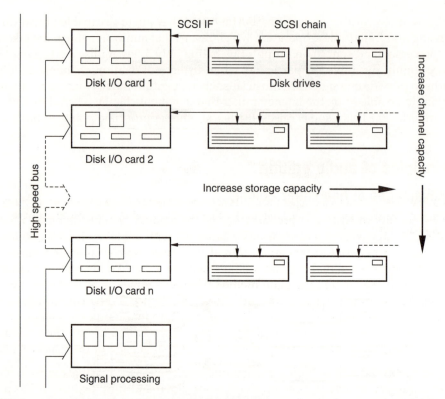

Figure 3.7 Arrangement of multiple disks in a typical modular system, showing how a number of disks can be attached to a single SCSI chain to increase storage capacity, and how additional disk I/O cards can be added to increase data throughput for additional audio channels

seamless join will result between old and new material, both at the start of the drop-in and at the drop-out into the old material again.

Dropping in and out are really very similar operations to those involved in editing, where a crossfade must be added between old and new material at the join. In terms of file operations, it may be appreciated that one cannot simply start to write new material half way through a previously written file, making it necessary to write a new file for the 'dropped-in' portion. Internally, as part of the replay schedule, the system will then have to keep a record of times at which it must crossfade from one file to the other and back again.

3.6 System latency

Latency is the delay that occurs between one event and another. In workstations the term latency is usually used to describe the delay between inputs and outputs of the audio hardware. It is particularly important because this latency affects the ease with which a workstation may be used as the principal audio signal-processing engine in a studio, this now being a realistic prospect. Large-scale audio workstations with multiple inputs and outputs can now handle most of the operations that would once have been handled by stand-alone mixers, effects and recording equipment. Furthermore they are now capable of real-time signal processing and 'full duplex' operation which means that audio signals can be taken from an input, processed and sent to an output, this taking only a few milliseconds. Low latency is therefore highly desirable, in particular when using workstation channels as foldback signal paths to provide cue signals to musicians when overdubbing new material. It may also be important to be able to fix the latency rather than having it change when different operations are undertaken. This issue is raised further at other appropriate points in the book.

3.7 Principles of audio editing

3.7.1 Advantages of non-linear editing

Speed and flexibility of editing is probably one of the greatest benefits obtained from non-linear recording. Tape editing had some advantages but with digital audio it was often cumbersome, requiring material to be copied in real time from source tapes to a master tape. Difficulties also arose when making minor adjustments to a finished master. Tape-cut editing was very fast and cheap, being the main method used for years with analog tape, but it was rather unreliable on digital formats and little used in practice. When cut-editing tape, the editor fixed the edited sections in a physical and therefore in a temporal relationship with each other. If he or she desired to change any aspect of the edited master then it would be taken apart and rejoined, there usually only being one final version of the master tape.

The majority of editing is done today using audio workstations. Non-linear editing has also come to feature very widely in post-production for video and film, because it has a lot in common with film post-production techniques involving a number of independent mono sound reels. The editor may preview a number of possible masters in their entirety before deciding which should be the final one. Even after this, it is a simple matter to modify the edit list to update the master. Edits may also be previewed and experimented with in order to determine the most appropriate location and processing – an operation which is less easy with other forms of editing.

This kind of editing is truly non-destructive because the edited master only exists as a series of instructions to replay parts of certain sound files at specified times, with optional signal processing overlaid, as shown in Figure 3.8. The original sound files remain intact at all times and a single sound file can be used as many times as desired in different locations and on different tracks without the need to duplicate the audio data. Editing may involve the simple joining of sections, or it may involve more complex operations such as long crossfades between one album track and the next, or gain offsets between one section and another. The beauty of non-linear editing is that all these things are possible without in any way affecting the original source material.

3.7.2 Sound files and sound segments

Sound files are discussed further in Chapter 6: they are the individual sound recordings contained on a disk, each of which is catalogued in the disk directory. In the case of music editing sound files might be session takes, anything from a few bars to a whole movement,

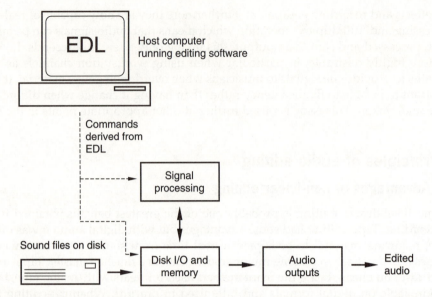

Figure 3.8 Instructions from an edit decision list (EDL) are used to control the replay of sound file segments from disk, which may be subjected to further processing (also under EDL control) before arriving at the audio outputs

while in picture dubbing they might contain a phrase of dialogue or a sound effect. They are normally stored with a name to identify them. Specific segments of these sound files can be defined by the user while editing, in order to get rid of unwanted material or to select useful extracts. In such cases it is useful to be able to identify the wanted segment as an entity in its own right, so that it can be named and used wherever required. The terminology varies but such identified parts of sound files are usually termed either 'clips' or 'segments'. They require the original sound files as source data and will not usually be replayable independently.

Rather than creating a copy of the segment or clip and storing it as a separate sound file, it is normal simply to store it as a 'soft' entity – in other words as simply commands in an edit list or project file that identify the start and end addresses of the segment concerned and the sound file to which it relates. It may be given a name by the operator and subsequently used as if it were a sound file in its own right. An almost unlimited number of these segments can be created from original sound files, without the need for any additional audio storage space.

3.7.3 Edit point handling

Edit points can be simple butt joins or crossfades. A butt join is very simple because it involves straightforward switching from the replay of one sound segment to another. Since replay involves temporary storage of the sound file blocks in RAM (see above) it is a relatively simple matter to ensure that both outgoing and incoming files in the region of the edit are available in RAM simultaneously (in different address areas). Up until the edit, blocks of the outgoing file are read from the disk into RAM and thence to the audio outputs. As the

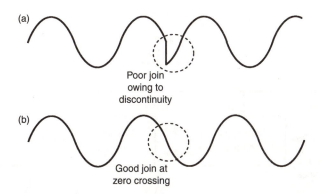

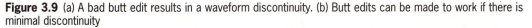

Figure 3.9 (a) A bad butt edit results in a waveform discontinuity. (b) Butt edits can be made to work if there is minimal discontinuity

edit point is reached a switch occurs between outgoing and incoming material by instituting a jump in the memory read address corresponding to the start of the incoming material. Replay then continues by reading subsequent blocks from the incoming sound file. It is normally possible to position edits right down to single sample accuracy, making the timing resolution as fine as a number of tens of microseconds if required.

The problem with butt joins is that they are quite unsubtle. Audible clicks and bumps may result because of the discontinuity in the waveform that may result, as shown in Figure 3.9. It is normal, therefore, to use at least a short crossfade at edit points to hide the effect of the join. This is what happens when analog tape is spliced, because the traditional angled cut has the same effect as a short crossfade (of between 5 and 20 ms depending on the tape speed and angle of cut). Most workstations have considerable flexibility with crossfades and are not limited to short durations. It is now common to use crossfades of many shapes and durations (e.g. linear, root cosine, equal power) for different creative purposes. This, coupled with the ability to preview edits and fine-tune their locations, has made it possible to put edits in places previously considered impossible.

The locations of edit points are kept in an edit decision list (EDL) which contains information about the segments and files to be replayed at each time, the in and the out points of each section and details of the crossfade time and shape at each edit point. It may also contain additional information such as signal processing operations to be performed (gain changes, EQ, etc.).

3.7.4 Crossfading

Crossfading is similar to butt joining, except that it requires access to data from both incoming and outgoing files for the duration of the crossfade. The crossfade calculation involves simple signal processing, during which the values of outgoing samples are multiplied by gradually decreasing coefficients whilst the values of incoming samples are multiplied by gradually increasing coefficients. Time coincident samples of the two files are then added together to produce output samples, as described in Chapter 2. The duration and shape of the crossfade

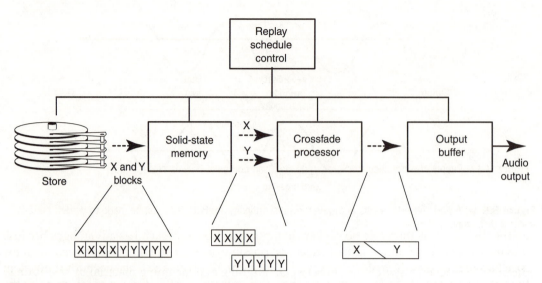

Figure 3.10 Conceptual diagram of the sequence of operations that occur during a crossfade. X and Y are the incoming and outgoing sound segments

can be adjusted by altering the coefficients involved and the rate at which the process is executed.

Crossfades are either performed in real time, as the edit point passes, or pre-calculated and written to disk as a file. There are merits to both approaches. Real-time crossfades can be varied at any time and are simply stored as commands in the EDL, indicating the nature of the fade to be executed. The process is similar to that for the butt edit, except that as the edit point approaches samples from both incoming and outgoing segments are loaded into RAM in order that there is an overlap in time. During the crossfade it is necessary to continue to load samples from both incoming and outgoing segments into their respective areas of RAM, and for these to be routed to the crossfade processor, as shown in Figure 3.10. The resulting samples are then available for routeing to the output. A consequence of this is that a temporary increase in disk activity occurs, because two streams of data rather than one are read during a crossfade. It is important, therefore, to have a disk drive and buffer size capable of handling the additional load present during real time crossfades, which represents a doubling in the transfer rate required. Eight channel replay would effectively become sixteen channel replay for the duration of a crossfade edit on all eight channels, for example. An editing system may consequently be pushed close to its limits if asked to perform long real time crossfades on multiple channels at the same time.

A common solution to this problem is for the crossfade to be calculated in non-real time when the edit point and crossfade duration is first determined by the user. This incurs a short delay while the system works out the sums, after which a new sound file is stored which is simply the crossfade period and nothing else. Replay of the edit is then a more simple matter, which involves playing the outgoing segment up to the beginning of the crossfade, then the crossfade file, then the incoming segment from after the crossfade, as shown in Figure 3.11. Load

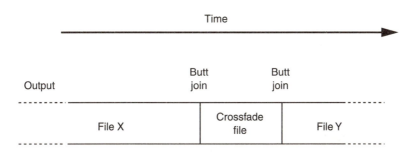

Figure 3.11 Replay of a precalculated crossfade file at an edit point between files X and Y

on the disk drive is therefore no higher than normal. This approach has advantages because it makes any number and length of crossfade possible on any combination of tracks, with the sure knowledge that they can be replayed. The slight disadvantage is the need for the system to write a new crossfade file every time the edit is altered and the disk space taken up by the crossfade files (although this is normally quite small).

The shape of the crossfade is often able to be changed to suit different operational purposes. Standard linear fades (those where the gain changes uniformly with time) are not always the most suitable for music editing, especially when the crossfade is longer than about ten milliseconds. The result may be a momentary drop in the resulting level in the centre of the crossfade that is due to the way in which the sound levels from the two files add together. If there is a random phase difference between the signals, as there will often be in music, the rise in level resulting from adding the two signals will normally be around 3 dB, but the linear crossfade is 6 dB down in its centre resulting in an overall level drop of around 3 dB (see Figure 3.12). Exponential crossfades and other such shapes may be more suitable for these purposes, because they have a smaller level drop in the centre. It may even be possible to design customised crossfade laws. Figure 3.13 shows the crossfade editing controls from a system by Sonic Solutions. It is possible to alter the offset of the start and end of the fade from the actual edit point and to have a faster fade up than fade down.

Many systems also allow automated gain changes to be introduced as well as fades, so that level differences across edit points may be corrected. Figure 3.14 shows a crossfade profile that has a higher level after the edit point than before it, and different slopes for the in and out fades. A lot of the difficulties that editors encounter in making edits work can be solved using a combination of these facilities.

3.7.5 Editing modes

During the editing process the operator will load appropriate sound files and audition them, both on their own and in a sequence with other files. The exact method of assembling the edited sequence depends very much on the user interface, but it is common to present the user with a visual analogy of moving tape, allowing files to be 'cut-and-spliced' or 'copied and pasted' into appropriate locations along the virtual tape. These files, or edited clips of

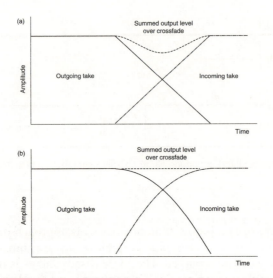

Figure 3.12 Summation of levels at a crossfade. (a) A linear crossfade can result in a level drop if the incoming and outgoing material are non-coherent. (b) An exponential fade, or other similar laws, can help to make the level more constant across the edit

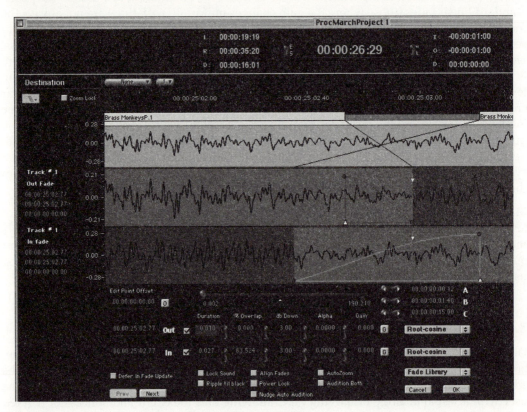

Figure 3.13 An example of crossfade control in Sonic Studio HD

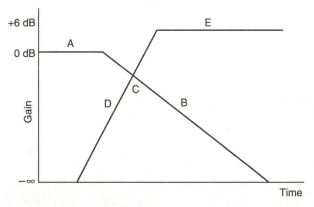

Figure 3.14 The system may allow the user to program a gain profile around an edit point, defining the starting gain (A), the fade-down time (B), the fade-up time (D), the point below unity at which the two files cross over (C) and the final gain (E)

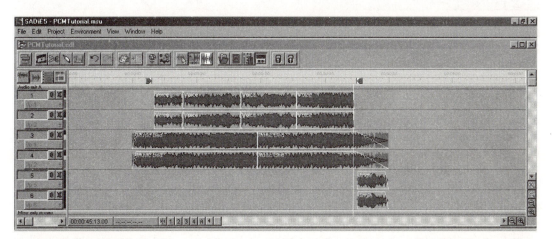

Figure 3.15 Example from SADiE editing system, showing audio clips assigned to different tracks on a virtual tape, against a timeline

them, are then played out at the timecode locations corresponding to their positions on this 'virtual tape' (an example is shown in Figure 3.15). It is also quite common to display a representation of the audio waveform that allows the editor to see as well as hear the signal around the edit point (see Figure 3.16).

In the editing of music using digital tape systems it was common to assemble an edited master from the beginning, copying takes from source tapes in sequence onto the master. An example of typical procedure will serve to illustrate the point. Starting at the beginning of the piece of music the first take would be copied to the master tape until a short time after the first edit was to be performed. The editor would then locate the edit point on the master tape (the outgoing take) by playing up to the approximate point and marking it, followed by fine

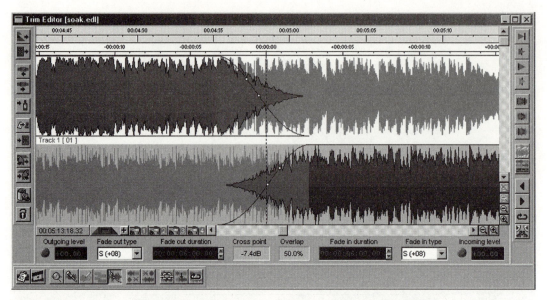

Figure 3.16 Example from SADiE editing system showing the 'trim editor' in which is displayed a detailed view of the audio waveform around the edit point, together with information about the crossfade

trimming of this point, either by nudging it in small time increments, or by the simulation of analog 'reel-rocking'. The edit point would then be confirmed and the same procedure performed on the source take to be joined at this point (the incoming take). This edit would then be auditioned, with a crossfade between outgoing and incoming material at the edit point, after which any further trimming would be performed before the edit was committed to the master tape by dropping it into record mode at the appropriate time.

In non-linear systems this approach is often simulated, allowing the user to roughly locate an edit point while playing the virtual tape followed by a fine trim using simulated reel-rocking or a detailed view of the waveform. Some software presents source and destination streams as well, in further simulation of the tape approach. Sound files and segments are treated as the equivalent of the 'takes' in the above example and the system notes the points in each segment at which one is to cease and another is to begin playing, with whatever overlap has been specified for cross-fading.

It is also possible to insert or change sections in the middle of a finished master, provided that the EDL and source files are still available. To take an example, assume that an edited opera has been completed and that the producer now wishes to change a take somewhere in the middle (see Figure 3.17). The replacement take is unlikely to be exactly the same length but it is possible simply to shuffle all of the following material along or back slightly to accommodate it, this being only a matter of changing the EDL rather than modifying the stored music in any way. The files are then simply played out at slightly different times than in the first version of the edit.

It is also normal to allow edited segments to be fixed in time if desired, so that they are not shuffled forwards or backwards when other segments are inserted. This 'anchoring' of

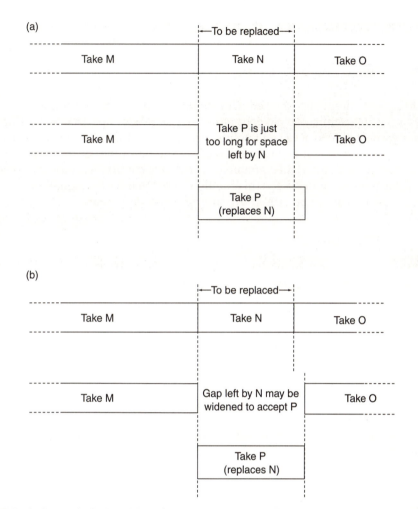

Figure 3.17 Replacing a take in the middle of an edited program. (a) Tape based copy editing results in a gap of fixed size, which may not match the new take length. (b) Non-linear editing allows the gap size to be adjusted to match the new take

segments is often used in picture dubbing when certain sound effects and dialogue have to remain locked to the picture.

3.7.6 Simulation of 'reel-rocking'

It is common to simulate the effect of reel-rocking in non-linear editors, providing the user with the sonic impression that reels of analog tape are being 'rocked' back and forth as they are in analog tape editing when fine-searching edit points. Editors are used to the sound of tape moving in this way, and are skilled at locating edit points when listening to such a sound.

The simulation of variable speed replay in both directions (forwards and backwards) is usually controlled by a wheel or sideways movement of a mouse which moves the 'tape' in either direction around the current play location. This magnitude and direction of this movement is used to control the rate at which samples are read from the disk file, via the buffer, and this replaces the fixed sampling rate clock as the controller of the replay rate. Systems differ very greatly as to the sound quality achieved in this mode, because it is in fact quite a difficult task to provide convincing simulation. So poor have been many attempts that many editors do not use the feature, preferring to judge edit points accurately 'on the fly', followed by trimming or nudging them either way if they are not successful the first time. Good simulation requires very fast, responsive action and an ergonomically suitable control. A mouse is very unsuitable for the purpose. It also requires a certain amount of DSP to filter the signal correctly, in order to avoid the aliasing that can be caused by varying the sampling rate.

4 MIDI and synthetic audio control

MIDI is the Music Instrument Digital Interface, a control protocol and interface standard for electronic musical instruments that has also been used widely in other music and audio products. Although it is relatively dated by modern standards it is still used extensively, which is something of a testament to its success. Even if the MIDI hardware interface is used less these days, either because more synthesis, sampling and processing takes place using software within the workstation, or because other data interfaces such as USB and Firewire are becoming popular, the protocol for communicating events and other control information is still widely encountered. A lot of software that runs on computers uses MIDI as a basis for controlling the generation of sounds and external devices.

Synthetic audio is used increasingly in audio workstations and mobile devices as a very efficient means of audio representation, because it only requires control information and sound object descriptions to be transmitted. Standards such as MPEG-4 Structured Audio enable synthetic audio to be used as an alternative or an addition to natural audio coding and this can be seen as a natural evolution of the MIDI concept in interactive multimedia applications.

4.1 Background

Electronic musical instruments existed widely before MIDI was developed in the early 1980s, but no universal means existed of controlling them remotely. Many older musical instruments used analogue voltage control, rather than being controlled by a microprocessor, and thus used a variety of analog remote interfaces (if indeed any facility of this kind was provided at all). Such interfaces commonly took the form of one port for timing information, such as might be required by a sequencer or drum machine, and another for pitch and key triggering information, as shown in Figure 4.1. The latter, commonly referred to as 'CV and gate', consisted of a DC (direct current) control line carrying a variable control voltage (CV)

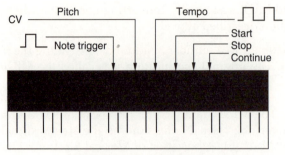

Electronic musical instrument

Figure 4.1 Prior to MIDI control, electronic musical instruments tended to use a DC remote interface for pitch and note triggering. A second interface handled a clock signal to control tempo and trigger pulses to control the execution of a stored sequence

which was proportional to the pitch of the note, and a separate line to carry a trigger pulse. A common increment for the CV was 1 volt per octave (although this was by no means the only approach) and notes on a synthesiser could be triggered remotely by setting the CV to the correct pitch and sending a 'note on' trigger pulse which would initiate a new cycle of the synthesiser's envelope generator. Such an interface would deal with only one note at a time, but many older synths were only monophonic in any case (that is, they were only capable of generating a single voice).

Instruments with onboard sequencers would need a timing reference in order that they could be run in synchronisation with other such devices, and this commonly took the form of a square pulse train at a rate related to the current musical tempo, often connected to the device using a DIN-type connector, along with trigger lines for starting and stopping a sequence's execution. There was no universal agreement over the rate of this external clock, and frequencies measured in pulses per musical quarter note (ppqn), such as 24 ppqn and 48 ppqn, were used by different manufacturers. A number of conversion boxes were available that divided or multiplied clock signals in order that devices from different manufacturers could be made to work together.

As microprocessor control began to be more widely used in musical instruments a number of incompatible digital control interfaces sprang up, promoted by the large synthesiser manufacturers, some serial and some parallel. Needless to say the plethora of non-standardised approaches to remote control made it difficult to construct an integrated system, especially when integrating equipment from different manufacturers. Owing to collaboration between the major parties in America and Japan, the way became cleared for agreement over a common hardware interface and command protocol, resulting in the specification of the MIDI standard in late 1982/early 1983. This interface grew out of an amalgamation of a proposed universal interface called USI (the Universal Synthesiser Interface) which was intended mainly for note on and off commands, and a Japanese specification which was rather more complex and which proposed an extensive protocol to cover other operations as well. Since MIDI's introduction, the use of older remote interfaces has died away very quickly, but there remain available a number of specialised interfaces which may be used to interconnect non-MIDI equipment to MIDI systems by converting the digital MIDI commands into the type of analog information described above.

The standard has been subject to a number of addenda, extending the functionality of MIDI far beyond the original. The original specification was called the MIDI 1.0 specification, to which has been added such addenda as the MIDI Sample Dump protocol, MIDI Files, General MIDI (1 and 2), MIDI TimeCode, MIDI Show Control, MIDI Machine Control and Downloadable Sounds. The MIDI Manufacturers Association (MMA) seems now to be the primary association governing formal extensions to the standard, liaising closely with a Japanese association called AMEI (Association of Musical Electronics Industry).

4.2 What is MIDI?

MIDI is a digital remote control interface for music systems. It follows that MIDI-controlled equipment is normally based on microprocessor control, with the MIDI interface forming an I/O port. It is a measure of the popularity of MIDI as a means of control that it has now been adopted in many other audio and visual systems, including the automation of mixing consoles, the control of studio outboard equipment, the control of lighting equipment and of other studio machinery. Although many of its standard commands are music related, it is possible either to adapt music commands to non-musical purposes or to use command sequences designed especially for alternative methods of control.

The adoption of a serial standard for MIDI was dictated largely by economic and practical considerations, as it was intended that it should be possible for the interface to be installed on relatively cheap items of equipment and that it should be available to as wide a range of users as possible. A parallel system might have been more professionally satisfactory, but would have involved a considerable manufacturing cost overhead per MIDI device, as well as parallel cabling between devices, which would have been more expensive and bulky than serial interconnection. The simplicity and ease of installation of MIDI systems has been largely responsible for its rapid proliferation as an international standard.

Unlike its analog predecessors, MIDI integrates timing and system control commands with pitch and note triggering commands, such that everything may be carried in the same format over the same piece of wire. MIDI makes it possible to control musical instruments polyphonically in pseudo real time: that is, the speed of transmission is such that delays in the transfer of performance commands are not audible in the majority of cases. It is also possible to address a number of separate receiving devices within a single MIDI data stream, and this allows a controlling device to determine the destination of a command.

4.3 MIDI and digital audio contrasted

For many the distinction between MIDI and digital audio may be a clear one, but those new to the subject often confuse the two. Any confusion is often due to both MIDI and digital audio equipment appearing to perform the same task – that is the recording of multiple channels of music using digital equipment – and is not helped by the way in which some manufacturers refer to MIDI sequencing as digital recording.

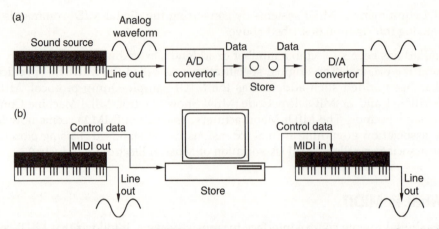

Figure 4.2 (a) Digital audio recording and (b) MIDI recording contrasted. In (a) the sound waveform itself is converted into digital data and stored, whereas in (b) only control information is stored, and a MIDI-controlled sound generator is required during replay

Digital audio involves a process whereby an audio waveform (such as the line output of a musical instrument) is sampled regularly and then converted into a series of binary words that represent the sound waveform, as described in Chapter 2. A digital audio recorder stores this sequence of data and can replay it by passing the original data through a digital-to-analog convertor that turns the data back into a sound waveform, as shown in Figure 4.2. A multitrack recorder has a number of independent channels that work in the same way, allowing a sound recording to be built up in layers. MIDI, on the other hand, handles digital information that *controls* the generation of sound. MIDI data does not represent the sound waveform itself. When a multitrack music recording is made using a MIDI sequencer (see Chapter 7) this control data is stored, and can be replayed by transmitting the original data to a collection of MIDI-controlled musical instruments. It is the instruments that actually reproduce the recording.

A digital audio recording, then, allows any sound to be stored and replayed without the need for additional hardware. It is useful for recording acoustic sounds such as voices, where MIDI is not a great deal of help. A MIDI recording is almost useless without a collection of sound generators. An interesting advantage of the MIDI recording is that, since the stored data represents event information describing a piece of music, it is possible to change the music by changing the event data. MIDI recordings also consume a lot less memory space than digital audio recordings. It is also possible to transmit a MIDI recording to a different collection of instruments from those used during the original recording, thus resulting in a different sound. It is now common for MIDI and digital audio recording to be integrated in one software package, allowing the two to be edited and manipulated in parallel.

4.4 Basic MIDI principles

4.4.1 System specifications

The MIDI hardware interface and connections are described in Chapter 5. MIDI is a serial interface, running at a relatively slow rate by modern standards, over which control

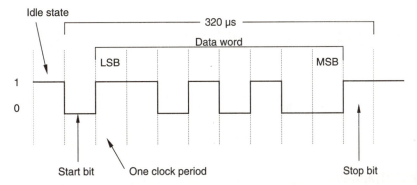

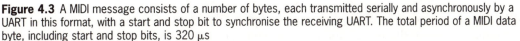

Figure 4.3 A MIDI message consists of a number of bytes, each transmitted serially and asynchronously by a UART in this format, with a start and stop bit to synchronise the receiving UART. The total period of a MIDI data byte, including start and stop bits, is 320 μs

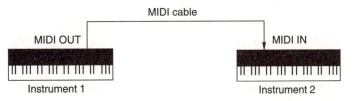

Figure 4.4 The simplest form of MIDI interconnection involves connecting two instruments together as shown

messages are sent as groups of bytes. Each byte is preceded by one start bit and followed by one stop bit per byte in order to synchronise reception of the data which is transmitted asynchronously, as shown in Figure 4.3. The addition of start and stop bits means that each 8-bit word actually takes 10 bit periods to transmit (lasting a total of 320 μs). Standard MIDI messages typically consist of one, two or three bytes, although there are longer messages for some purposes that will be covered later in this book.

4.4.2 Simple interconnection

In the simplest MIDI system, one instrument could be connected to another as shown in Figure 4.4. Here, instrument 1 sends information relating to actions performed on its own controls (notes pressed, pedals pressed, etc.) to instrument 2, which imitates these actions as far as it is able. This type of arrangement can be used for 'doubling-up' sounds, 'layering' or 'stacking', such that a composite sound can be made up from two synthesisers' outputs. (The audio outputs of the two instruments would have to be mixed together for this effect to be heard.) Larger MIDI systems could be built up by further 'daisy-chaining' of instruments, such that instruments further down the chain all received information generated by the first (see Figure 4.5), although this is not a very satisfactory way of building a large MIDI system. In large systems some form of central routing helps to avoid MIDI 'traffic jams' and simplifies interconnection.

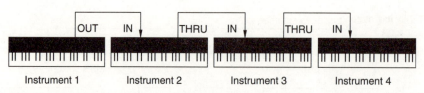

Instrument 1   Instrument 2   Instrument 3   Instrument 4

Figure 4.5 Further instruments can be added using THRU ports as shown, in order that messages from instrument 1 may be transmitted to all the other instruments

4.4.3 MIDI channels

MIDI messages are made up of a number of bytes. Each part of the message has a specific purpose, and one of these is to define the receiving channel to which the message refers. In this way, a controlling device can make data device-specific – in other words it can define which receiving instrument will act on the data sent. This is most important in large systems that use a computer sequencer as a master controller, when a large amount of information will be present on the MIDI data bus, not all of which is intended for every instrument. If a device is set in software to receive on a specific channel or on a number of channels it will act only on information which is 'tagged' with its own channel numbers. Everything else it will usually ignore. There are 16 basic MIDI channels and instruments can usually be set to receive on any specific channel or channels (*omni off* mode), or to receive on all channels (*omni on* mode). The latter mode is useful as a means of determining whether anything at all is being received by the device.

Later it will be seen that the limit of 16 MIDI channels can be overcome easily by using multiport MIDI interfaces connected to a computer. In such cases it is important not to confuse the MIDI data channel with the physical port to which a device may be connected, since each physical port will be capable of transmitting on all 16 data channels.

4.4.4 Message format

There are two basic types of MIDI message byte: the status byte and the data byte. The first byte in a MIDI message is normally a status byte. Standard MIDI messages can be up to three bytes long, but not all messages require three bytes, and there are some fairly common exceptions to the rule which are described below. Table 4.1 shows the format and content of MIDI messages under each of the statuses.

Status bytes always begin with a binary one to distinguish them from data bytes, which always begin with a zero. Because the most significant bit (MSB) of each byte is reserved to denote the type (status or data) there are only seven active bits per byte which allows 2^7 (that is 128) possible values. As shown in Figure 4.6, the first half of the status byte denotes the message type and the second half denotes the channel number. Because four bits of the status byte are set aside to indicate the channel number, this allows for 2^4 (or 16) possible channels. There are only three bits to denote the message type, because the first bit must always be a one. This theoretically allows for eight message types, but there are some special cases in the form of system messages (see below).

Table 4.1 MIDI messages summarised

Message	Status	Data 1	Data 2
Note off	&8n	Note number	Velocity
Note on	&9n	Note number	Velocity
Polyphonic aftertouch	&An	Note number	Pressure
Control change	&Bn	Controller number	Data
Program change	&Cn	Program number	–
Channel aftertouch	&Dn	Pressure	–
Pitch wheel	&En	LSbyte	MSbyte
System exclusive			
System exclusive start	&F0	Manufacturer ID	Data, (Data), (Data)
End of SysEx	&F7	–	
System common			
Quarter frame	&F1	Data	–
Song pointer	&F2	LSbyte	MSbyte
Song select	&F3	Song number	–
Tune request	&F6	–	
System realtime			
Timing clock	&F8	–	–
Start	&FA	–	–
Continue	&FB	–	–
Stop	&FC	–	–
Active sensing	&FE	–	–
Reset	&FF	–	–

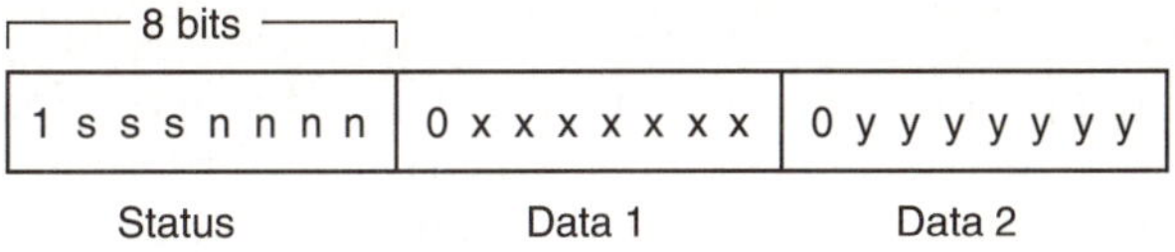

```
        ┌──── 8 bits ────┐
        ┌───────────────┬───────────────┬───────────────┐
        │ 1 s s s n n n n │ 0 x x x x x x x │ 0 y y y y y y y │
        └───────────────┴───────────────┴───────────────┘
             Status           Data 1           Data 2
```

Figure 4.6 General format of a MIDI message. The 'sss' bits are used to define the message type, the 'nnnn' bits define the channel number, whilst the 'xxxxxxx' and 'yyyyyyy' bits carry the message data. See text for details

4.5 MIDI messages in detail

In this section the MIDI communication protocol will be examined in detail. The majority of the basic message types and their meanings will be explained. The descriptions here are not intended as an alternative to reading the MIDI documentation itself, but rather as a commentary on it and an explanation of it. It follows that examples will be given, but that the reader should refer to the standard for a full description of the protocol. The standard has been extended and refined over the years and the following is to be regarded as an introduction to the basic messages. The prefix '&' will be used to indicate hexadecimal values throughout the discussion; individual MIDI message bytes will be delineated using square brackets, e.g. [&45], and channel numbers will be denoted using 'n' to indicate that the value may be anything from &0 to &F (channels 1 to 16).

The MMA has defined Approved Protocols (APs) and Recommended Practices (RPs). An AP is a part of the standard MIDI specification and is used when the standard is further defined or when a previously undefined command is defined, whereas an RP is used to describe an optional new MIDI application that is not a mandatory or binding part of the standard. Not all MIDI devices will have all the following commands implemented, since it is not mandatory for a device conforming to the MIDI standard to implement every possibility.

4.5.1 Channel and system messages contrasted

Two primary classes of message exist: those that relate to specific MIDI channels and those that relate to the system as a whole. One should bear in mind that it is possible for an instrument to be receiving in 'omni on' mode, in which case it will ignore the channel label and attempt to respond to anything that it receives.

Channel messages start with status bytes in the range &8n to &En (they start at hexadecimal eight because the MSB must be a one for a status byte). System messages all begin with &F, and do not contain a channel number. Instead the least significant nibble of the system status byte is used for further identification of the system message, such that there is room for 16 possible system messages running from &F0 to &FF. System messages are themselves split into three groups: system common, system exclusive and system realtime. The common messages may apply to any device on the MIDI bus, depending only on the device's ability to handle the message. The exclusive messages apply to whichever manufacturer's devices are specified later in the message (see below) and the realtime messages are intended for devices which are to be synchronised to the prevailing musical tempo. (Some of the so-called realtime messages do not really seem to deserve this appellation, as discussed below.) The status byte &F1 is used for MIDI TimeCode.

MIDI channel numbers are usually referred to as 'channels one to sixteen', but it can be appreciated that in fact the binary numbers that represent these run from zero to fifteen (&0 to &F), as fifteen is the largest decimal number which can be represented with four bits. Thus the note on message for channel 5 is actually &94 (nine for note on, and four for channel 5).

4.5.2 Note on and note off messages

Much of the musical information sent over a typical MIDI interface will consist of these two message types. As indicated by the titles, the note on message turns on a musical note, and the note off message turns it off. Note on takes the general format:

```
[&8n] [Note number] [Velocity]
```

and note off takes the form:

```
[&9n] [Note number] [Velocity]  (although see Section 4.5.3)
```

A MIDI instrument will generate note on messages at its MIDI OUT corresponding to whatever notes are pressed on the keyboard, on whatever channel the instrument is set to transmit. Also, any note which has been turned on must subsequently be turned off in order for it

Table 4.2 MIDI note numbers related to the musical scale

Musical note	MIDI note number
C–2	0
C–1	12
C0	24
C1	36
C2	48
C3 (middle C)	60 (Yamaha convention)
C4	72
C5	84
C6	96
C7	108
C8	120
G8	127

to stop sounding, thus if one instrument receives a note on message from another and then loses the MIDI connection for any reason, the note will continue sounding *ad infinitum*. This situation can occur if a MIDI cable is pulled out during transmission.

MIDI note numbers relate directly to the western musical chromatic scale and the format of the message allows for 128 note numbers which cover a range of a little over ten octaves – adequate for the full range of most musical material. This quantisation of the pitch scale is geared very much towards keyboard instruments, being less suitable for other instruments and cultures where the definition of pitches is not so black and white. Nonetheless, means have been developed of adapting control to situations where unconventional tunings are required. Note numbers normally relate to the musical scale as shown in Table 4.2, although there is a certain degree of confusion here. Yamaha established the use of C3 for middle C, whereas others have used C4. Some software allows the user to decide which convention will be used for display purposes.

4.5.3 Velocity information

Note messages are associated with a velocity byte that is used to represent the speed at which a key was pressed or released. The former will correspond to the force exerted on the key as it is depressed: in other words, 'how hard you hit it' (called 'note on velocity'). It is used to control parameters such as the volume or timbre of the note at the audio output of an instrument and can be applied internally to scale the effect of one or more of the envelope generators in a synthesiser. This velocity value has 128 possible states, but not all MIDI instruments are able to generate or interpret the velocity byte, in which case they will set it to a value half way between the limits, i.e.: 64_{10}. Some instruments may act on velocity information even if they are unable to generate it themselves. It is recommended that a logarithmic rather than linear relationship should be established between the velocity value and the parameter which it controls, since this corresponds more closely to the way in which musicians expect an instrument to respond, although some instruments allow customised mapping of velocity values to parameters. The note on, velocity zero value is reserved for the special purpose of turning a note off, for reasons which will become clear in Section 4.5.4. If an instrument sees a note number with a velocity of zero, its software should interpret this as a note off message.

Note off velocity (or 'release velocity') is not widely used, as it relates to the speed at which a note is released, which is not a parameter that affects the sound of many normal keyboard instruments. Nonetheless it is available for special effects if a manufacturer decides to implement it.

4.5.4 Running status

Running status is an accepted method of reducing the amount of data transmitted. It involves the assumption that once a status byte has been asserted by a controller there is no need to reiterate this status for each subsequent message of that status, so long as the status has not changed in between. Thus a string of note on messages could be sent with the note on status only sent at the start of the series of note data, for example:

```
[&9n] [Data] [Velocity] [Data] [Velocity] [Data] [Velocity]
```

For a long string of notes this could reduce the amount of data sent by nearly one third. But in most music each note on is almost always followed quickly by a note off for the same note number, so this method would clearly break down as the status would be changing from note on to note off very regularly, thus eliminating most of the advantage gained by running status. This is the reason for the adoption of note on, velocity zero as equivalent to a note off message, because it allows a string of what appears to be note on messages, but which is, in fact, both note on and note off.

Running status is not used at all times for a string of same-status messages and will often only be called upon by an instrument's software when the rate of data exceeds a certain point. Indeed, an examination of the data from a typical synthesiser indicates that running status is not used during a large amount of ordinary playing.

4.5.5 Polyphonic key pressure (aftertouch)

The key pressure messages are sometimes called 'aftertouch' by keyboard manufacturers. Aftertouch is perhaps a slightly misleading term as it does not make clear what aspect of touch is referred to, and many people have confused it with note off velocity. This message refers to the amount of pressure placed on a key at the bottom of its travel, and it is used to instigate effects based on how much the player leans onto the key after depressing it. It is often applied to performance parameters such as vibrato.

The polyphonic key pressure message is not widely used, as it transmits a separate value for every key on the keyboard and thus requires a separate sensor for every key. This can be expensive to implement and is beyond the scope of many keyboards, so most manufacturers have resorted to the use of the channel pressure message (see below). The message takes the general format:

```
[&An] [Note number] [Pressure]
```

Implementing polyphonic key pressure messages involves the transmission of a considerable amount of data that might be unnecessary, as the message will be sent for every note in a chord every time the pressure changes. As most people do not maintain a constant pressure on the bottom of a key whilst playing, many redundant messages might be sent per note. A technique known as 'controller thinning' may be used by a device to limit the rate at which such messages are transmitted and this may be implemented either before transmission or at a later stage using a computer. Alternatively this data may be filtered out altogether if it is not required.

4.5.6 Control change

As well as note information, a MIDI device may be capable of transmitting control information that corresponds to the various switches, control wheels and pedals associated with it. These come under the control change message group and should be distinguished from program change messages. The controller messages have proliferated enormously since the early days of MIDI and not all devices will implement all of them. The control change message takes the general form:

```
[&Bn] [Controller number] [Data]
```

so a number of controllers may be addressed using the same type of status byte by changing the controller number.

Although the original MIDI standard did not lay down any hard and fast rules for the assignment of physical control devices to logical controller numbers, there is now common agreement amongst manufacturers that certain controller numbers will be used for certain purposes. These are assigned by the MMA. There are two distinct kinds of controller: the switch type and the analog type. The analog controller is any continuously variable wheel, lever, slider or pedal that might have any one of a number of positions and these are often known as continuous controllers. There are 128 controller numbers available and these are grouped as shown in Table 4.3. Table 4.4 shows a more detailed breakdown of some of these, as found in the majority of MIDI-controlled musical instruments, although the full list is regularly updated by the MMA. The control change messages have become fairly complex and interested users are referred to the relevant standards.

Table 4.3 MIDI controller classifications

Controller number (hex)	Function
&00–1F	14 bit controllers, MSbyte
&20–3F	14 bit controllers, LSbyte
&40–65	7 bit controllers or switches
&66–77	Originally undefined
&78–7F	Channel mode control

Table 4.4 MIDI controller functions

Controller number (hex)	Function
00	Bank select
01	Modulation wheel
02	Breath controller
03	Undefined
04	Foot controller
05	Portamento time
06	Data entry slider
07	Main volume
08	Balance
09	Undefined
0A	Pan
0B	Expression controller
0C	Effect control 1
0D	Effect control 2
0E–0F	Undefined
10–13	General purpose controllers 1–4
14–1F	Undefined
20–3F	LSbyte for 14 bit controllers (same function order as 00–1F)
40	Sustain pedal
41	Portamento on/off
42	Sostenuto pedal
43	Soft pedal
44	Legato footswitch
45	Hold 2
46–4F	Sound controllers
50–53	General purpose controllers 5–8
54	Portamento control
55–5A	Undefined
5B–5F	Effects depth 1–5
60	Data increment
61	Data decrement
62	NRPC LSbyte (non-registered parameter controller)
63	NRPC MSbyte
64	RPC LSbyte (registered parameter controller)
65	RPC MSbyte
66–77	Undefined
78	All sounds off
79	Reset all controllers
7A	Local on/off
7B	All notes off
7C	Omni receive mode off
7D	Omni receive mode on
7E	Mono receive mode
7F	Poly receive mode

The first 64 controller numbers (that is up to &3F) relate to only 32 *physical* controllers (the continuous controllers). This is to allow for greater resolution in the quantisation of position than would be feasible with the seven bits that are offered by a single data byte. Seven bits would only allow 128 possible positions of an analog controller to be represented and this might not be adequate in some cases. For this reason the first 32 controllers handle the most significant byte (MSbyte) of the controller data, while the second 32 handle the least significant byte (LSbyte). In this way, controller numbers &06 and &38 both represent the data entry slider, for example. Together, the data values can make up a 14-bit number (because the first bit of each data word has to be a zero), which allows the quantisation of a control's position to be one part in 2^{14} (16384_{10}). Clearly, not all controllers will require this resolution, but it is available if needed. Only the LSbyte would be needed for small movements of a control. If a system opts not to use the extra resolution offered by the second byte, it should send only the MSbyte for coarse control. In practice this is all that is transmitted on many devices.

On/off switches can be represented easily in binary form (0 for OFF, 1 for ON), and it would be possible to use just a single bit for this purpose, but, in order to conform to the standard format of the message, switch states are normally represented by data values between &00 and &3F for OFF and &40–&7F for ON. In other words switches are now considered as 7-bit continuous controllers. In older systems it may be found that only &00 = OFF and &7F = ON.

The data increment and decrement buttons that are present on many devices are assigned to two specific controller numbers (&60 and &61) and an extension to the standard defines four controllers (&62 to &65) that effectively expand the scope of the control change messages. These are the registered and non-registered parameter controllers (RPCs and NRPCs).

The 'all notes off' command (frequently abbreviated to 'ANO') was designed to be transmitted to devices as a means of silencing them, but it does not necessarily have this effect in practice. What actually happens varies between instruments, especially if the sustain pedal is held down or notes are still being pressed manually by a player. All notes off is supposed to put all note generators into the release phase of their envelopes, and clearly the result of this will depend on what a sound is programmed to do at this point. The exception should be notes which are being played while the sustain pedal is held down, which should only be released when that pedal is released. 'All sounds off' was designed to overcome the problems with 'all notes off', by turning sounds off as quickly as possible. 'Reset all controllers' is designed to reset all controllers to their default state, in order to return a device to its 'standard' setting.

4.5.7 Channel modes

Although grouped with the controllers, under the same status, the channel mode messages differ somewhat in that they set the mode of operation of the instrument receiving on that particular channel.

'Local on/off' is used to make or break the link between an instrument's keyboard and its own sound generators. Effectively there is a switch between the output of the keyboard and the control input to the sound generators which allows the instrument to play its own sound generators in normal operation when the switch is closed (see Figure 4.7). If the switch is opened, the link is broken and the output from the keyboard feeds the MIDI OUT while the

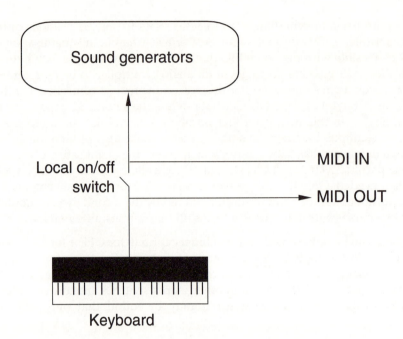

Figure 4.7 The 'local off' switch disconnects a keyboard from its associated sound generators in order that the two parts may be treated independently in a MIDI system

sound generators are controlled from the MIDI IN. In this mode the instrument acts as two separate devices: a keyboard without any sound, and a sound generator without a keyboard. This configuration can be useful when the instrument in use is the master keyboard for a large sequencer system, where it may not always be desired that everything played on the master keyboard results in sound from the instrument itself.

'Omni off' ensures that the instrument will only act on data tagged with its own channel number(s), as set by the instrument's controls. 'Omni on' sets the instrument to receive on all of the MIDI channels. In other words, the instrument will ignore the channel number in the status byte and will attempt to act on any data that may arrive, whatever its channel. Devices should power-up in this mode according to the original specification, but more recent devices will tend to power up in the mode that they were left. Mono mode sets the instrument such that it will only reproduce one note at a time, as opposed to 'Poly' (phonic) in which a number of notes may be sounded together.

In older devices the mono mode came into its own as a means of operating an instrument in a 'multitimbral' fashion, whereby MIDI information on each channel controlled a separate monophonic musical voice. This used to be one of the only ways of getting a device to generate more than one type of voice at a time. The data byte that accompanies the mono mode message specifies how many voices are to be assigned to adjacent MIDI channels, starting with the basic receive channel. For example, if the data byte is set to 4, then four voices will be assigned to adjacent MIDI channels, starting from the basic channel which is the one on which the instrument has been set to receive in normal operation. Exceptionally, if the data byte is set to 0, all 16 voices (if they exist) are assigned each to one of the 16 MIDI channels. In this way, a single multitimbral instrument can act as 16 monophonic instruments, although on cheaper systems all of these voices may be combined to one audio output.

Mono mode tends to be used mostly on MIDI guitar synthesisers because each string can then have its own channel and each can control its own set of pitch bend and other parameters. The mode also has the advantage that it is possible to play in a truly legato fashion – that is with a smooth take over between the notes of a melody – because the arrival of a second note message acts simply to change the pitch if the first one is still being held down, rather than re-triggering the start of a note envelope. The legato switch controller (see Table 2.4) allows a similar type of playing in polyphonic modes by allowing new note messages only to change the pitch.

In poly mode the instrument will sound as many notes as it is able at the same time. Instruments differ as to the action taken when the number of simultaneous notes is exceeded: some will release the first note played in favour of the new note, whereas others will refuse to play the new note. Some may be able to route excess note messages to their MIDI OUT ports so that they can be played by a chained device. The more intelligent of them may look to see if the same note already exists in the notes currently sounding and only accept a new note if is not already sounding. Even more intelligently, some devices may release the quietest note (that with the lowest velocity value), or the note furthest through its velocity envelope, to make way for a later arrival. It is also common to run a device in poly mode on more than one receive channel, provided that the software can handle the reception of multiple polyphonic channels. A multitimbral sound generator may well have this facility, commonly referred to as 'multi' mode, making it act as if it were a number of separate instruments each receiving on a separate channel. In multi mode a device may be able to dynamically assign its polyphony between the channels and voices in order that the user does not need to assign a fixed polyphony to each voice.

4.5.8 Program change

The program change message is used most commonly to change the 'patch' of an instrument or other device. A patch is a stored configuration of the device, describing the setup of the tone generators in a synthesiser and the way in which they are interconnected. Program change is channel-specific and there is only a single data byte associated with it, specifying to which of 128 possible stored programs the receiving device should switch. On non-musical devices such as effects units, the program change message is often used to switch between different effects and the different effects programs may be mapped to specific program change numbers. The message takes the general form:

```
&[Cn] [Program number]
```

If a program change message is sent to a musical device it will usually result in a change of voice, as long as this facility is enabled. Exactly which voice corresponds to which program change number depends on the manufacturer. It is quite common for some manufacturers to implement this function in such a way that a data value of zero gives voice number one. This results in a permanent offset between the program change number and the voice number, which should be taken into account in any software. On some instruments, voices may be split into a number of 'banks' of 8, 16 or 32, and higher banks can be selected over MIDI by setting the program change number to a value which is 8, 16 or 32 higher than the lowest bank number. For example, bank 1, voice 2, might be selected by program change &01, whereas bank 2, voice 2, would probably be selected in this case by program change &11, where there were 16 voices per bank.

There are also a number of other approaches used in commercial sound modules. Where more than 128 voices need to be addressed remotely, the more recent 'bank select' command may be implemented.

4.5.9 Channel aftertouch

Most instruments use a single sensor, often in the form of a pressure-sensitive conductive plastic bar running the length of the keyboard, to detect the pressure applied to keys at the bottom of their travel. In the case of channel aftertouch, one message is sent for the entire instrument and this will correspond to an approximate total of the pressure over the range of the keyboard, the strongest influence being from the key pressed the hardest. (Some manufacturers have split the pressure detector into upper and lower keyboard regions, and some use 'intelligent' zoning.) The message takes the general form:

```
&[Dn] [Pressure value]
```

There is only one data byte, so there are 128 possible values and, as with the polyphonic version, many messages may be sent as the pressure is varied at the bottom of a key's travel. Controller 'thinning' may be used to reduce the quantity of these messages, as described above.

4.5.10 Pitch bend wheel

The pitch wheel message has a status byte of its own, and carries information about the movement of the sprung-return control wheel on many keyboards which modifies the pitch of any note(s) played. It uses two data bytes in order to give 14 bits of resolution, in much the same way as the continuous controllers, except that the pitch wheel message carries both bytes together. Fourteen data bits are required so that the pitch appears to change smoothly, rather than in steps (as it might with only seven bits). The pitch bend message is channel specific so ought to be sent separately for each individual channel. This becomes important when using a single multi-timbral device in mono mode (see above), as one must ensure that a pitch bend message only affects the notes on the intended channel. The message takes the general form:

```
&[En] [LSbyte] [MSbyte]
```

The value of the pitch bend controller should be halfway between the lower and upper range limits when it is at rest in its sprung central position, thus allowing bending both down and up. This corresponds to a hex value of &2000, transmitted as &[En] [00] [40]. The range of pitch controlled by the bend message is set on the receiving device itself, or using the RPC designated for this purpose (see Section 4.6.7).

4.5.11 System exclusive

A system exclusive message is one that is unique to a particular manufacturer and often a particular instrument. The only thing that is defined about such messages is how they are to start and finish, with the exception of the use of system exclusive messages for universal information, as discussed elsewhere. System exclusive messages generated by a device will naturally be produced at the MIDI OUT, not at the THRU, so a deliberate connection must be made between the transmitting device and the receiving device before data transfer may take place. Occasionally it is necessary to make a return link from the OUT of the receiver to the IN of the transmitter so that two-way communication is possible and so that the receiver can control the flow of data to some extent by telling the transmitter when it is ready to receive and when it has received correctly (a form of handshaking).

The message takes the general form:

```
&[F0] [ident.] [data] [data] ... [F7]
```

where [ident.] identifies the relevant manufacturer ID, a number defining which manufacturer's message is to follow. Originally, manufacturer IDs were a single byte but the number of IDs has been extended by setting aside the [00] value of the ID to indicate that two further bytes of ID follow. Manufacturer IDs are therefore either one or three bytes long. A full list of manufacturer IDs is available from the MMA.

Data of virtually any sort can follow the ID. It can be used for a variety of miscellaneous purposes that have not been defined in the MIDI standard and the message can have virtually any length that the manufacturer requires. It is often split into packets of a manageable size in order not to cause receiver memory buffers to overflow. Exceptions are data bytes that look like other MIDI status bytes (except realtime messages), as they will naturally be interpreted as such by any receiver, which might terminate reception of the system exclusive message. The message should be terminated with &F7, although this is not always observed, in which case the receiving device should 'time-out' after a given period, or terminate the system exclusive message on receipt of the next status byte. It is recommended that some form of error checking (typically a checksum) is employed for long system exclusive data dumps, and many systems employ means of detecting whether the data has been received accurately, asking for re-tries of sections of the message in the event of failure, via a return link to the transmitter.

Examples of applications for such messages can be seen in the form of sample data dumps (from a sampler to a computer and back again for editing purposes), although this is painfully slow, and voice data dumps (from a synthesiser to a computer for storage and editing of user-programmed voices). There are now an enormous number of uses of system exclusive messages, both in the universal categories and in the manufacturer categories.

4.5.12 Universal system exclusive messages

The three highest numbered IDs within the system exclusive message have been set aside to denote special modes. These are the 'universal non-commercial' messages (ID: &7D), the 'universal non-realtime' messages (ID: &7E) and the 'universal realtime' messages (ID: &7F). Universal sysex messages are often used for controlling device parameters that were not originally specified in the MIDI standard and that now need addressing in most devices. Examples are things like 'chorus modulation depth', 'reverb type' and 'master fine tuning'.

Universal non-commercial messages are set aside for educational and research purposes and should not be used in commercial products. Universal non-realtime messages are used for universal system exclusive events which are not time critical and universal realtime messages deal with time critical events (thus being given a higher priority). The two latter types of message normally take the general form of:

```
&[F0] [ID] [dev. ID] [sub-ID #1] [sub-ID #2] [data] ... ... [F7]
```

Device ID used to be referred to as 'channel number', but this did not really make sense since a whole byte allows for the addressing of 128 channels and this does not correspond to the normal 16 channels of MIDI. The term 'device ID' is now used widely in software as a means of defining one of a number of physical devices in a large MIDI system, rather than defining a MIDI channel number. It should be noted, though, that it is allowable for a device to have more than one ID if this seems appropriate. Modern MIDI devices will normally allow their device ID to be set either over MIDI or from the front panel. The use of &7F in this position signifies that the message applies to all devices as opposed to just one.

The sub-IDs are used to identify firstly the category or application of the message (sub-ID #1) and secondly the type of message within that category (sub-ID #2). For some reason, the original MIDI sample dump messages do not use the sub-ID #2, although some recent additions to the sample dump do.

4.5.13 Tune request

Older analog synthesisers tended to drift somewhat in pitch over the time that they were turned on. The tune request is a request for these synthesisers to re-tune themselves to a fixed reference. (It is advisable not to transmit pitch bend or note on messages to instruments during a tune up because of the unpredictable behaviour of some products under these conditions.)

4.5.14 Active sensing

Active sensing messages are single status bytes sent roughly three times per second by a controlling device when there is no other activity on the bus. They act as a means of reassuring the receiving devices that the controller has not disappeared. Not all devices transmit active sensing information, and a receiver's software should be able to detect the presence or lack of it. If a receiver has come to expect active sensing bytes then it will generally act by turning off all notes if these bytes disappear for any reason. This can be a useful function when a MIDI cable has been pulled out during a transmission, as it ensures that notes will not be left sounding for very long. If a receiver has not seen active sensing bytes since last turned on, it should assume that they are not being used.

4.5.15 Reset

This message resets all devices on the bus to their power-on state. The process may take some time and some devices mute their audio outputs, which can result in clicks, therefore the message should be used with care.

4.6 MIDI control of sound generators

4.6.1 MIDI note assignment in synthesisers and samplers

Many of the replay and signal processing aspects of synthesis and sampling now overlap so that it is more difficult to distinguish between the two. In basic terms a sampler is a device that stores short clips of sound data in RAM, enabling them to be replayed subsequently at different pitches, possibly looped and processed. A synthesiser is a device that enables signals to be artificially generated and modified to create novel sounds. Wavetable synthesis is based on a similar principle to sampling, though, and stored samples can form the basis for synthesis. A sound generator can often generate a number of different sounds at the same time. It is possible that these sounds could be entirely unrelated (perhaps a single drum, an animal noise and a piano note), or that they might have some relationship to each other (perhaps a number of drums in a kit, or a selection of notes from a grand piano). The method by which sounds or samples are assigned to MIDI notes and channels is defined by the replay program.

The most common approach when assigning note numbers to samples is to program the sampler with the range of MIDI note numbers over which a certain sample should be sounded. Akai, one of the most popular sampler manufacturers, calls these 'keygroups'. It may be that this 'range' is only one note, in which case the sample in question would be triggered only on receipt of that note number, but in the case of a range of notes the sample would be played on receipt of any note in the range. In the latter case transposition would be required, depending on the relationship between the note number received and the original note number given to the sample (see above). A couple of examples highlight the difference in approach, as shown in Figure 4.8. In the first example, illustrating a possible approach to note assignment for a collection of drum kit sounds, most samples are assigned to only one note number, although it is possible for tuned drum sounds such as tom-toms to be assigned over a range in order to give the impression of 'tuned toms'. Each MIDI note message received would replay the particular percussion sound assigned to that note number in this example.

In the second example, illustrating a suggested approach to note assignment for an organ, notes were originally sampled every musical fifth across the organ's note range. The replay program has been designed so that each of these samples is assigned to a note range of a fifth, centred on the original pitch of each sample, resulting in a maximum transposition of a third up or down. Ideally, of course, every note would have been sampled and assigned to an individual note number on replay, but this requires very large amounts of memory and painstaking sample acquisition in the first place.

In further pursuit of sonic accuracy, some devices provide the facility for introducing a crossfade between note ranges. This is used where an abrupt change in the sound at the boundary between two note ranges might be undesirable, allowing the takeover from one sample

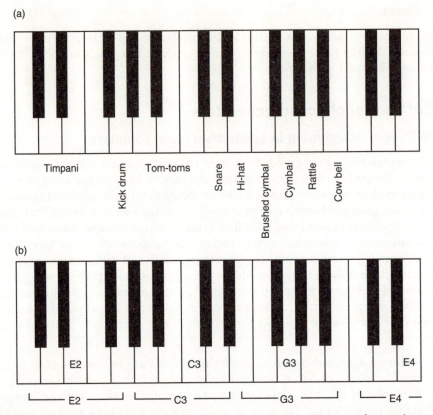

Figure 4.8 (a) Percussion samples are often assigned to one note per sample, except for tuned percussion which sometimes covers a range of notes. (b) Organ samples could be transposed over a range of notes, centred on the original pitch of the sample

to another to be more gradual. For example, in the organ scenario introduced above, the timbre could change noticeably when playing musical passages that crossed between two note ranges because replay would switch from the upper limit of transposition of one sample to the lower limit of the next (or vice versa). In this case the ranges for the different samples are made to overlap (as illustrated in Figure 4.9). In the overlap range the system mixes a proportion of the two samples together to form the output. The exact proportion depends on the range of overlap and the note's position within this range. Very accurate tuning of the original samples is needed in order to avoid beats when using positional crossfades. Clearly this approach would be of less value when each note was assigned to a completely different sound, as in the drum kit example.

Crossfades based on note velocity allow two or more samples to be assigned to one note or range of notes. This requires at least a 'loud sample' and a 'soft sample' to be stored for each original sound and some systems may accommodate four or more to be assigned over the

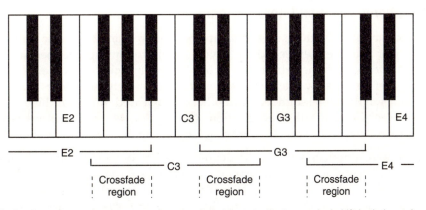

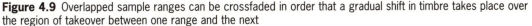

Figure 4.9 Overlapped sample ranges can be crossfaded in order that a gradual shift in timbre takes place over the region of takeover between one range and the next

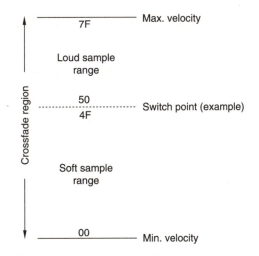

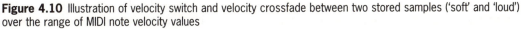

Figure 4.10 Illustration of velocity switch and velocity crossfade between two stored samples ('soft' and 'loud') over the range of MIDI note velocity values

velocity range. The terminology may vary, but the principle is that a velocity value is set at which the replay switches from one stored sample to another, as many instruments sound quite different when they are loud to when they are soft (it is more than just the volume that changes: it is the timbre also). If a simple switching point is set, then the change from one sample to the other will be abrupt as the velocity crosses either side of the relevant value. This can be illustrated by storing two completely different sounds as the loud and soft samples, in which case the output changes from one to the other at the switching point. A more subtle effect is achieved by using velocity crossfading, in which the proportion of loud and soft samples varies depending on the received note velocity value. At low velocity values the proportion of the soft sample in the output would be greatest and at high values the output content would be almost entirely made up of the loud sample (see Figure 4.10).

4.6.2 Polyphony, voice and note assignment

Modern sound modules (synthesisers and samplers) tend to be multi-note polyphonic. When the polyphony of a device is exceeded the device should follow a predefined set of rules to determine what to do with the extra notes. Typically a sound module will either release the 'oldest' notes first, or possibly release the quietest. Alternatively, new notes that exceed the polyphony will simply not be sounded until others are released. Rules for this are defined in some of the recent General MIDI specifications (see Section 4.8), and composers may now even be able to exercise some control over what happens in devices with limited polyphony.

It is important to distinguish between the degree of polyphony offered by a device and the number of simultaneous *voices* it can generate. Sometimes these may be traded off against each other in multi-timbral devices, by allocating a certain number of notes to each voice, with the total adding up to the total polyphony. Either 16 notes could be allocated to one voice or four notes to each of four voices, for example. Dynamic allocation is often used to distribute the polyphony around the voices depending on demand and this is a particular feature of General MIDI sound modules.

A multi-timbral sound generator is one that is capable of generating more than one voice at a time, independent of polyphony considerations. A voice is a particular sound type, such as 'grand piano' or 'accordion'. This capability is now the norm for modern sound modules. Older synthesisers used to be able to generate only one or two voices at a time, possibly allowing a keyboard split, and could sometimes make use of MIDI channel mode 4 (monophonic, omni off) to allow multiple *monophonic* voices to be generated under MIDI control. They tended only to receive polyphonically on one MIDI channel at a time. More recent systems are capable of receiving on all 16 MIDI channels simultaneously, with each channel controlling an entirely independent polyphonic voice.

4.6.3 MIDI functions of sound generators

The MIDI implementation for a particular sound generator should be described in the manual that accompanies it. A MIDI implementation chart such as the one shown in Figure 4.11 indicates which message types are received and transmitted, together with any comments relating to limitations or unusual features. Functions such as note off velocity and polyphonic aftertouch, for example, are quite rare. It is quite common for a device to be able to accept certain data and act upon it, even if it cannot generate such data from its own controllers. The note range available under MIDI control compared with that available from a device's keyboard is a good example of this, since many devices will respond to note data over a full ten octave range yet still have only a limited (or no) keyboard. This approach can be used by a manufacturer who wishes to make a cheaper synthesiser that omits the expensive physical sensors for such things as velocity and aftertouch, while retaining these functions in software for use under MIDI control. Devices conforming to the General MIDI specification described in Section 4.8 must conform to certain basic guidelines concerning their MIDI implementation and the structure of their sound generators.

4.6.4 MIDI data buffers and latency

All MIDI-controlled equipment uses some form of data buffering for received MIDI messages. Such buffering acts as a temporary store for messages that have arrived but have not

YAMAHA (Tone Generator) Model TG100 MIDI Implementation Chart			Version : 1.00
Function	Transmitted	Recognised	Remarks
Basic Default	X	1–16	Memorised
Channel Changed	X	1–16	
Mode Default	X	3	
Messages	X	3,4 (m = 1) *2	
Altered	**************	X	
Note	X	O - 127	
Number : True voice	**************	O - 127	
Velocity Note ON	X	O 9nH, v=1–127	
Note OFF	X	X	
After Key's	X	X	
Touch Ch's	X	O	
Pitch Bender	X	O 0–24 semi	12-bit resolution
Control 0,32	X	O MSB only	Bank select
1	X	O	Modulation wheel
5	X	O	Portamento time
6,38	X	O	Data entry
7	X	O *1	Volume
Change 10	X	O	Panpot
11	X	O *1	Expression
64	X	O	Hold 1
65	X	O	Portamento
91	X	O (Reverb)	Effect depth 1
100,101	X	O	PRN LSB, MSB
120	X	O	All sound off
121	X	O	Reset all controls
Prog	X	O 0–127 *1	
Change : True #	**************		
System Exclusive	O *3	O *3	
System : Song Pos.	X	X	
: Song Sel.	X	X	
Common : Tune	X	X	
System : Clock	X	X	
Real Time: Commands	X	X	
Aux : Local ON/OFF	X	X	
: All Notes OFF	X	O (123–127)	
Messages: Active Sense	X	O	
: Reset	X	X	

Notes :   *1   ; receive if switch is on.
         *2   ; m is always treated as "1" regardless of its value.
         *3   ; transmit /receive if exclusive switch is on.

Mode 1 : OMNI ON, POLY	Mode 2 : OMNI ON, POLY	O :Yes
Mode 3 : OMNI OFF, POLY	Mode 4 : OMNI OFF, MONO	X : No

Figure 4.11 A typical MIDI implementation chart for a synthesiser sound module. (Yamaha TG100, with permission)

yet been processed and allows for a certain prioritisation in the handling of received messages. Cheaper devices tend to have relatively small MIDI input buffers and these can overflow easily unless care is taken in the filtering and distribution of MIDI data around a large system (usually accomplished by a MIDI router or multiport interface). When a buffer overflows it will normally result in an error message displayed on the front panel of the device, indicating that some MIDI data is likely to have been lost. More advanced equipment can store more MIDI data in its input buffer, although this is not necessarily desirable because many messages that are transmitted over MIDI are intended for 'real-time' execution and one would not wish them to be delayed in a temporary buffer. Such buffer delay is one potential cause of latency in MIDI systems. A more useful solution would be to speed up the rate at which incoming messages are processed.

4.6.5 Handling of velocity and aftertouch data

Sound generators able to respond to note on velocity will use the value of this byte to control assigned functions within the sound generators. It is common for the user to be able to program the device such that the velocity value affects certain parameters to a greater or lesser extent. For example, it might be decided that the 'brightness' of the sound should increase with greater key velocity, in which case it would be necessary to program the device so that the envelope generator that affected the brightness was subject to control by the velocity value. This would usually mean that the maximum effect of the envelope generator would be limited by the velocity value, such that it could only reach its full programmed effect (that which it would give if not subject to velocity control) if the velocity was also maximum. The exact law of this relationship is up to the manufacturer and may be used to simulate different types of 'keyboard touch'. A device may offer a number of laws or curves relating changes in velocity to changes in the control value, or the received velocity value may be used to scale the preset parameter rather than replace it.

Another common application of velocity value is to control the amplitude envelope of a particular sound, such that the output volume depends on how hard the key is hit. In many synthesiser systems that use multiple interacting digital oscillators, these velocity-sensitive effects can all be achieved by applying velocity control to the envelope generator of one or more of the oscillators, as indicated earlier in this chapter.

Note off velocity is not implemented in many keyboards, and most musicians are not used to thinking about what they do as they release a key, but this parameter can be used to control such factors as the release time of the note or the duration of a reverberation effect. Aftertouch (either polyphonic or channel, as described in Section 4.5) is often used in synthesisers to control the application of low frequency modulation (tremolo or vibrato) to a note. Sometimes aftertouch may be applied to other parameters, but this is less common.

4.6.6 Handling of controller messages

The controller messages that begin with a status of &Bn, as listed in Table 4.4, turn up in various forms in sound generator implementations. It should be noted that although there are standard definitions for many of these controller numbers it is often possible to remap them either within sequencer software or within sound modules themselves. Fourteen-bit continuous controllers are rarely encountered for any parameter and often only the MSbyte of the controller value (which uses the first 32 controller numbers) is sent and used. For most parameters the 128 increments that result are adequate.

Controllers &07 (Volume) and &0A (Pan) are particularly useful with sound modules as a means of controlling the internal mixing of voices. These controllers work on a per channel basis, and are independent of any velocity control which may be related to note volume. There are two real-time system exclusive controllers that handle similar functions to these, but for the device as a whole rather than for individual voices or channels. The 'master volume' and 'master balance' controls are accessed using:

```
&[F0] [7F] [dev. ID] [04] [01 or 02] [data] [data] [F7]
```

where the sub-ID #1 of &04 represents a 'device control' message and sub-ID #2s of &01 or &02 select volume or balance respectively. The [data] values allow 14-bit resolution for the parameters concerned, transmitted LSB first. Balance is different to pan because pan sets the stereo positioning (the split in level between left and right) of a mono source, whereas balance sets the relative levels of the left and right channels of a stereo source (see Figure 4.12). Since a pan or balance control is used to shift the stereo image either left or right from a centre detent position, the MIDI data values representing the setting are ranged either side of a mid-range value that corresponds to the centre detent. The channel pan controller is thus

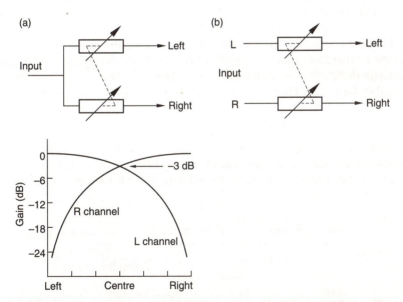

Figure 4.12 (a) A pan control takes a mono input and splits it two ways (left and right), the stereo position depending on the level difference between the two channels. The attenuation law of pan controls is designed to result in a smooth movement of the source across the stereo 'picture' between left and right, with no apparent rise or fall in overall level when the control is altered. A typical pan control gain law is shown here. (b) A balance control simply adjusts the relative level between the two channels of a stereo signal so as to shift the entire stereo image either left or right

normally centred at a data value of 63 (and sometimes over a range of values just below this if the pan has only a limited number of steps), assuming that only a single 7-bit controller value is sent. There may be fewer steps in these controls than there are values of the MIDI controller, depending on the device in question, resulting in a range of controller values that will give rise to the same setting.

Some manufacturers have developed alternative means of expressive control for synthesisers such as the 'breath controller', which is a device which responds to the blowing effort applied by the mouth of the player. It was intended to allow wind players to have more control over expression in performance. Plugged into the synthesiser, it can be applied to various envelope generator or modulator parameters to affect the sound. The breath controller also has its own MIDI controller number. There is also a portamento controller (&54) that defines a note number from which the next note should slide. It is normally transmitted between two note on messages to create an automatic legato portamento effect between two notes.

The 'effects' and 'sound' controllers have been set aside as a form of general purpose control over aspects of the built-in effects and sound quality of a device. How they are applied will depend considerably on the architecture of the sound module and the method of synthesis used, but they give some means by which a manufacturer can provide a more abstracted form of control over the sound without the user needing to know precisely which voice parameters to alter. In this way, a user who is not prepared to get into the increasingly complicated world of voice programming can modify sounds to some extent.

The effects controllers occupy five controller numbers from &5B to &5F and are defined as Effects Depths 1–5. The default names for the effects to be controlled by these messages are respectively 'External Effects Depth', 'Tremolo Depth', 'Chorus Depth', 'Celeste (Detune) Depth' and 'Phaser Depth', although these definitions are open to interpretation and change by manufacturers. There are also ten sound controllers that occupy controller numbers from &46 to &4F. Again these are user- or manufacturer-definable, but five defaults were originally specified (listed in Table 4.5). They are principally intended as real-time controllers to be used during performance, rather than as a means of editing internal voice patches (the RPCs and NRPCs can be used for this as described below).

The sound variation controller is interesting because it is designed to allow the selection of one of a number of variants on a basic sound, depending on the data value that follows the controller number. For example, a piano sound might have variants of 'honky tonk', 'soft

Table 4.5 Sound controller functions (byte 2 of status &Bn)

MIDI controller number	Function (default)
&46	Sound variation
&47	Timbre/harmonic content
&48	Release time
&49	Attack time
&4A	Brightness
&4B–4F	No default

pedal', 'lid open' and 'lid closed'. The data value in the message is not intended to act as a continuous controller for certain voice parameters, rather the different data values possible in the message are intended to be used to select certain pre-programmed variations on the voice patch. If there are less than the 128 possible variants on the voice then the variants should be spread evenly over the number range so that there is an equal number range between them.

The timbre and brightness controllers can be used to alter the spectral content of the sound. The timbre controller is intended to be used specifically for altering the harmonic content of a sound, whilst the brightness controller is designed to control its high frequency content. The envelope controllers can be used to modify the attack and release times of certain envelope generators within a synthesiser. Data values less than &40 attached to these messages should result in progressively shorter times, whilst values greater than &40 should result in progressively longer times.

4.6.7 Registered and non-registered parameter numbers

The MIDI standard was extended a few years ago to allow for the control of individual internal parameters of sound generators by using a specific control change message. This meant, for example, that any aspect of a voice, such as the velocity sensitivity of an envelope generator, could be assigned a parameter number that could then be accessed over MIDI and its setting changed, making external editing of voices much easier. Parameter controllers are a subset of the control change message group, and they are divided into the registered and non-registered numbers (RPNs and NRPNs). RPNs are intended to apply universally and should be registered with the MMA, whilst NRPNs may be manufacturer specific. Only five parameter numbers were originally registered as RPNs, as shown in Table 4.6, but more may be added at any time and readers are advised to check the most recent revisions of the MIDI standard.

Parameter controllers operate by specifying the address of the parameter to be modified, followed by a control change message to increment or decrement the setting concerned. It is also possible to use the data entry slider controller to alter the setting of the parameter. The address of the parameter is set in two stages, with an MSbyte and then an LSbyte message, so as to allow for 16 384 possible parameter addresses. The controller numbers &62 and &63 are used to set the LS- and MSbytes respectively of an NRPN, whilst &64 and &65 are used

Table 4.6 Some examples of RPC definitions

RPC number (hex)	Parameter
00 00	Pitch bend sensitivity
00 01	Fine tuning
00 02	Coarse tuning
00 03	Tuning program select
00 04	Tuning bank select
7F 7F	Cancels RPN or NRPN (usually follows Message 3)

to address RPNs. The sequence of messages required to modify a parameter is as follows:

Message 1

```
&[Bn] [62 or 64] [LSB]
```

Message 2

```
&[Bn] [63 or 65] [MSB]
```

Message 3

```
&[Bn] [60 or 61] [7F] or &[Bn] [06] [DATA] [38] [DATA]
```

Message 3 represents either data increment (&60) or decrement (&61), or a 14-bit data entry slider control change with MSbyte (&06) and LSbyte (&38) parts (assuming running status). If the control has not moved very far, it is possible that only the MSbyte message need be sent.

4.6.8 Voice selection

The program change message was adequate for a number of years as a means of selecting one of a number of stored voice patches on a sound generator. Program change on its own allows for up to 128 different voices to be selected and a synthesiser or sound module may allow a program change map to be set up in order that the user may decide which voice is selected on receipt of a particular message. This can be particularly useful when the module has more than 128 voices available, but no other means of selecting voice banks. A number of different program change maps could be stored, perhaps to be selected under system exclusive control.

Modern sound modules tend to have very large patch memories – often too large to be adequately addressed by 128 program change messages. Although some older synthesisers used various odd ways of providing access to further banks of voices, most modern modules have implemented the standard 'bank select' approach. In basic terms, 'bank select' is a means of extending the number of voices that may be addressed by preceding a standard program change message with a message to define the bank from which that program is to be recalled. It uses a 14-bit control change message, with controller numbers &00 and &20, to form a 14-bit bank address, allowing 16 384 banks to be addressed. The bank number is followed directly by a program change message, thus creating the following general message:

```
&[Bn] [00] [MSbyte (of bank)]
&[Bn] [20] [LSbyte]
&[Cn] [Program number]
```

4.7 MIDI tuning control

Conventional equal-tempered tuning is the norm in western musical environments, but there may be cases when alternative tuning standards are required in order to conform to other

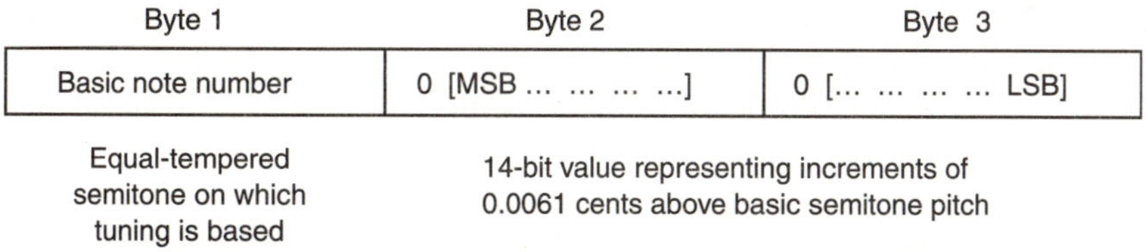

Byte 1	Byte 2	Byte 3
Basic note number	0 [MSB ... ...]	0 [... ... ... LSB]

| Equal-tempered semitone on which tuning is based | 14-bit value representing increments of 0.0061 cents above basic semitone pitch | |

Figure 4.13 MIDI tuning messages indicate the pitches to which MIDI notes should be tuned using three bytes, as shown here

temperaments or to non-western musical styles. Many devices now have the capability to store a number of alternative tuning maps, or to be retuned 'on the fly'. A number of manu-facturer-specific methods were used in the past (prior to the MIDI Tuning Standard), being SysEx messages preceded by the relevant manufacturer ID, but the MIDI Tuning Standard now forms the basis for communicating information about alternative tunings.

The tuning standard assumes that any note on a sound generator can be tuned over the entire range 8.1758 Hz to 13 289.73 Hz. It then allows individual notes' tuning to be adjusted in frac-tions of a semitone above a conventional MIDI note's pitch (which would be based on the equal temperament convention). A semitone is divided into 100 cents. A cent is one hun-dredth of a semitone, and as such does not represent a constant frequency increment in hertz but represents a proportion of the frequency of the note concerned. As the pitch of the basic note rises, so the frequency increment represented by a cent also increases. Two MIDI data bytes are used to indicate the fraction of a semitone above the basic note pitch, so the maxi-mum resolution possible is 100 cents/2^{14} which equals 0.0061 cents.

Tuning of individual notes is represented by three MIDI messages in total. The first specifies a numbered semitone in the MIDI note range on which the fractional tuning is to be based (the same as the MIDI note number in a note on message) and the second and third form a pair containing a 14-bit number (the MSB of each is 0), transmitted MSB first. This 14-bit number is used as described in the previous paragraph, with each increment representing a change of 0.0061 cents upwards in pitch from the basic semitone number (see Figure 4.13). A sound generator that is not capable of tuning to the accuracy contained in the message should tune to the nearest possible value, but it is recommended that it *stores* the full resolu-tion tuning value in tuning memories, in case data is to be transmitted to other devices which are capable of full resolution. The frequency value of &[7F] [7F] [7F] is reserved to indicate no change to the tuning of a particular note.

A number of MIDI messages are associated with tuning. These break down into bulk dumps of tuning data (to retune a complete instrument), single note retuning messages and the selection of prestored tuning programs and banks of programs. The only one of these that is currently a real-time message is the single note retuning. Users may select stored tuning programs and banks of tuning programs using the RPN messages shown in Table 4.6. A device may request a bulk tuning dump from another using the general SysEx non-realtime form:

```
&[F0]  [7E]  [dev. ID]  [08]  [00]  [tt]  [F7]
```

where the sub-ID #1 of &08 indicates a MIDI tuning standard message and the sub-ID #2 of &00 indicates a bulk dump request. &[tt] defines the tuning program which is being requested. Such a request should result in the transmission of a bulk dump if such a tuning program exists, and the dump should take the form:

```
&[F0] [7E] [dev. ID] [08] [01] [tt] [tuning name] ... ... [tuning data] ...
... ... [LL] [F7]
```

where [tuning name] is 16 bytes to name the tuning program (each byte holds a 7-bit ASCII character) and [tuning data] consists of 128 groups of 3 bytes to define the tuning of each note, in the format described in the previous section. &LL is a checksum byte.

A single note may be retuned using the SysEx realtime message:

```
&[F0] [7F] [dev. ID] [08] [02] [tt] [ll] ([kk] [tuning data]) ... ... [F7]
```

where &[ll] indicates the number of notes to be retuned, followed by that number of groups of tuning data. Each group of tuning data is preceded by &[kk] which defines the note to be retuned.

4.8 General MIDI

One of the problems with MIDI sound generators is that although voice patches can be selected using MIDI program change commands, there is no guarantee that a particular program change number will recall a particular voice on more than one instrument. In other words, program change 3 may correspond to 'alto sax' on one instrument and 'grand piano' on another. This makes it difficult to exchange songs between systems with any hope of the replay sounding the same as intended by the composer. General MIDI is an approach to the standardisation of a sound generator's behaviour, so that songs can be exchanged more easily between systems and device behaviour can be predicted by controllers. It comes in three flavours: GM 1, GM Lite and GM 2.

General MIDI Level 1 specifies a standard voice map and it specifies a minimum degree of polyphony, requiring that a sound generator should be able to receive MIDI data on all 16 channels simultaneously and polyphonically, with a different voice on each channel. There is also a requirement that the sound generator should support percussion sounds in the form of drum kits, so that a General MIDI sound module is capable of acting as a complete 'band in a box'.

Dynamic voice allocation is the norm in GM sound modules, with a requirement either for at least 24 dynamically allocated voices in total, or 16 for melody and 8 for percussion. Voices should all be velocity sensitive and should respond at least to the controller messages 1, 7, 10, 11, 64, 121 and 123 (decimal), RPNs 0, 1 and 2 (see above), pitch bend and channel aftertouch. In order to ensure compatibility between sequences that are replayed on GM modules, percussion sounds are always allocated to MIDI channel 10. Program change numbers are mapped to specific voice names, with ranges of numbers allocated to certain types of sounds, as shown in Table 4.7. Precise voice names may be found in the GM documentation. Channel 10,

Table 4.7 General MIDI program number ranges (except channel 10)

Program change (decimal)	Sound type
0–7	Piano
8–15	Chromatic percussion
16–23	Organ
24–31	Guitar
32–39	Bass
40–47	Strings
48–55	Ensemble
56–63	Brass
64–71	Reed
72–79	Pipe
80–87	Synth lead
88–95	Synth pad
96–103	Synth effects
104–111	Ethnic
112–119	Percussive
120–128	Sound effects

the percussion channel, has a defined set of note numbers on which particular sounds are to occur, so that the composer may know for example that key 39 will always be a 'hand clap'.

General MIDI sound modules may operate in modes other than GM, where voice allocations may be different, and there are two universal non-realtime SysEx messages used to turn GM on or off. These are:

```
&[F0] [7E] [dev. ID] [09] [01] [F7]
```

to turn GM on, and:

```
&[F0] [7E] [dev. ID] [09] [02] [F7]
```

to turn it off.

There is some disagreement over the definition of 'voice', as in '24 dynamically allocated voices' – the requirement that dictates the degree of polyphony supplied by a GM module. The spirit of the GM specification suggests that 24 notes should be capable of sounding simultaneously, but some modules combine sound generators to create composite voices, thereby reducing the degree of note polyphony.

General MIDI Lite (GML) is a cut-down GM 1 specification designed mainly for use on mobile devices with limited processing power. It can be used for things like ring tones on mobile phones and for basic music replay from PDAs. It specifies a fixed polyphony of 16 simultaneous notes, with 15 melodic instruments and 1 percussion kit on channel 10. The voice map is the same as GM Level 1. It also supports basic control change messages and the pitch-bend sensitivity RPN. As a rule, GM Level 1 songs will usually replay on GM Lite devices with acceptable quality, although some information may not be reproduced. An alternative to GM Lite is SPMIDI (see Section 4.9) which allows greater flexibility.

GM Level 2 is backwards compatible with Level 1 (GM 1 songs will replay correctly on GM 2 devices) but allows the selection of voice banks and extends polyphony to 32 voices. Percussion kits can run on channel 11 as well as the original channel 10. It adds MIDI tuning, RPN controllers and a range of universal system exclusive messages to the MIDI specification, enabling a wider range of control and greater versatility.

4.9 Scalable polyphonic MIDI (SPMIDI)

SPMIDI, rather like GM Lite, is designed principally for mobile devices that have issues with battery life and processing power. It has been adopted by the 3GPP wireless standards body for structured audio control of synthetic sounds in ring tones and multimedia messaging. It was developed primarily by Nokia and Beatnik. The SPMIDI basic specification for a device is based on GM Level 2, but a number of selectable profiles are possible, with different levels of sophistication.

The idea is that rather than fixing the polyphony at 16 voices the polyphony should be scalable according to the device profile (a description of the current capabilities of the device). SPMIDI also allows the content creator to decide what should happen when polyphony is limited – for example, what should happen when only four voices are available instead of 16. Conventional 'note stealing' approaches work by stealing notes from sounding voices to supply newly arrived notes, and the outcome of this can be somewhat arbitrary. In SPMIDI this is made more controllable. A process known as channel masking is used, whereby certain channels have a higher priority than others, enabling the content creator to put high priority material on particular channels. The channel priority order and maximum instantaneous polyphony are signalled to the device in a setup message at the initialisation stage.

4.10 Standard MIDI files (SMF)

Sequencers and notation packages typically store data on disk in their own unique file formats. The standard MIDI file was developed in an attempt to make interchange of information between packages more straightforward and it is now used widely in the industry in addition to manufacturers' own file formats. It is rare now not to find a sequencer or notation package capable of importing and exporting standard MIDI files. MIDI files are most useful for the interchange of performance and control information. They are not so useful for music notation where it is necessary to communicate greater detail about the way music appears on the stave and other notational concepts. For the latter purpose a number of different file formats have been developed, including Music XML which is among the most widely used of the universal interchange formats today. Further information about Music XML resources and other notation formats may be found in the Further reading at the end of this chapter.

Three types of standard MIDI file exist to encourage the interchange of sequencer data between software packages. The MIDI file contains data representing events on individual sequencer tracks, as well as containing labels such as track names, instrument names and time signatures.

4.10.1 General structure of MIDI files

There are three MIDI file types. File type 0 is the simplest and is used for single-track data, whilst file type 1 supports multiple tracks which are 'vertically' synchronous with each other (such as the parts of a song). File type 2 contains multiple tracks that have no direct timing relationship and may therefore be asynchronous. Type 2 could be used for transferring song files made up of a number of discrete sequences, each with a multiple track structure.

The basic file format consists of a number of 8-bit words formed into chunk-like parts, very similar to the RIFF and AIFF audio file formats described in Chapter 6. SMFs are not exactly RIFF files though, because they do not contain the highest level FORM chunk. (To encapsulate SMFs in a RIFF structure, use the RMID format, described in Section 4.12.) The header chunk, which always heads a MIDI file, contains global information relating to the whole file, whilst subsequent track chunks contain event data and labels relating to individual sequencer tracks. Track data should be distinguished from MIDI channel data, since a sequencer track may address more than one MIDI channel. Each chunk is preceded by a preamble of its own, which specifies the type of chunk (header or track) and the length of the chunk in terms of the number of data bytes that are contained in the chunk. There then follow the designated number of data bytes (see Figure 4.14). The chunk preamble contains 4 bytes to identify the chunk type using ASCII representation and 4 bytes to indicate the number of data bytes in the chunk (the length). The number of bytes indicated in the length does not include the preamble (which is always 8 bytes).

4.10.2 Header chunk

The header chunk takes the format shown in Figure 4.15. After the 8-byte preamble will normally be found 6 bytes containing header data, considered as three 16-bit words, the first of

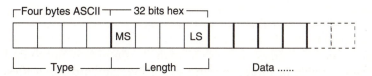

Figure 4.14 The general format of a MIDI file chunk. Each chunk has a preamble consisting of a 4-byte ASCII 'type' followed by 4 bytes to represent the number of data bytes in the rest of the message (the 'length')

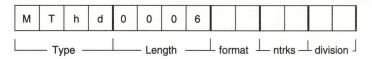

Figure 4.15 The header chunk has the type 'MThd' and the number of data bytes indicated in the 'length' is 6 (see text)

which ('format') defines the file type as 0, 1 or 2 (see above), the second of which ('ntrks') defines the number of track chunks in the file, and the third of which ('division') defines the timing format used in subsequent track events.

A zero in the MSB of the 'division' word indicates that events will be represented by 'musical' time increments of a certain number of 'ticks per quarter note' (the exact number is defined in the remaining bits of the word), whilst a one in the MSB indicates that events will be represented by real-time increments in number-of-ticks-per-timecode-frame. The frame rate of the timecode is given in the remaining bits of the most significant byte of 'division', being represented using negative values in twos complement form, so the standard frame rates are represented by one of the decimal values -24, -25, -29 (for 30 drop frame) or -30.

When a real-time format is specified in the header chunk, the least significant byte of 'division' is used to specify the subdivisions of a frame to which events may be timed. For example, a value of '4_{10}' in this position would mean that events were timed to an accuracy of a quarter of a frame, corresponding to the arrival frequency of MIDI quarter-frame timecode messages, whilst a value of '80_{10}' would allow events to be timed to bit accuracy within the timecode frame (there are 80 bits representing a single timecode frame value in the SMPTE/EBU longitudinal timecode format).

4.10.3 Track chunks

Following the header come a number of track chunks (see Figure 4.16), the number depending on the file type and the number of tracks. File type 0 represents a single track and will only contain a header and one track chunk, whilst file types 1 and 2 may have many track chunks. Track chunks contain strings of MIDI events, each labelled with a delta-time at which the event is to occur. Delta-times represent the number of 'ticks' since the last event, as opposed to the absolute time since the beginning of a song. The exact time increment specified by a tick depends on the definition of a tick contained in the 'division' word of the header (see above).

Delta-time values are represented in 'variable length format', which is a means of representing hexadecimal numbers up to &0FFFFFFF as compactly as possible. Variable length values

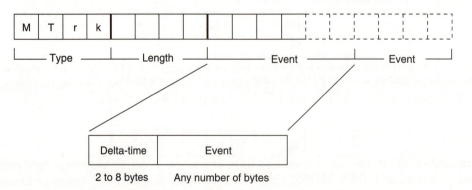

Figure 4.16 A track chunk has the type 'MTrk' and the number of data bytes indicated in the 'length' depends on the contents of the chunk. The data bytes which follow are grouped into events as shown

Table 4.8 Examples of numbers in variable length format

Original number (hex)	Variable length format (hex)
00000000	00
00000040	40
0000007F	7F
00000080	81 00
00002000	C0 00
00100000	C0 80 00
0FFFFFFF	FF FF FF 7F

represent the number in question using one, two, three or four bytes, depending on the size of the number. Each byte of the variable length value has its MSB set to a one, except for the last byte whose MSB should be zero. (This distinguishes the last byte of the value from the others, so that the computer reading the data knows when to stop compiling the number.) Seven bits of each byte are therefore available for the representation of numeric data (rather like the MIDI status and data bytes). A software routine must be written to convert normal hex values into this format and back again. The standard document gives some examples of hex numbers and their variable length equivalents, as shown in Table 4.8.

4.10.4 MIDI file track events

The track events that occur at specified delta-times fall into the categories of 'MIDI event', 'SysEx event' and 'meta-event'. In the case of the MIDI event, the data bytes that follow the delta-time are simply those of a MIDI channel message, with running status used if possible.

System exclusive (SysEx) events are used for holding MIDI system exclusive dumps that occur during a sequence. The event data is normally identical to the system exclusive data packet to be transmitted, except that the length of the packet is specified *after* the initial &[F0] byte that signals the beginning of a SysEx message and before the normal manufacturer ID, as follows:

```
&[F0] [length] [SysEx data] ... ...
```

The 'length' value should be encoded in variable length format, and the standard requires that &[F7] be used to terminate a SysEx event in a MIDI file. (Some software omits this when transmitting such data over MIDI.) It is also possible to have a special 'SysEx' event, as follows:

```
&[F7] [length] [data] ... ...
```

The standard says that this can be used as a form of 'escape' event, in order that data may be included in a standard MIDI file that would not normally be part of a sequencer file, such as real-time messages or MTC messages. The &F7 byte is also used as an identifier for subsequent parts of a system exclusive message that is to be transmitted in timed packets (some instruments require this). In such a case the first packet of the SysEx message uses the &F0 identifier and subsequent packets use the &F7 identifier, preceded by the appropriate delta-times to ensure correct timing of the packets.

The meta-event is used for information such as time signature, key signature, text, lyrics, instrument names and tempo markings. Its general format consists of a delta-time followed by the identifier &FF, as follows:

```
&[FF] [type] [length] [data] ... ...
```

The byte following &FF defines the type of meta-event, and the 'length' value is a variable length number describing the number of data bytes in the message which follows it. The number of bytes taken up by 'length' therefore depends on the message length to be represented.

Many meta-events exist and it is not intended to describe them all here, although some of the most common type identifiers are listed in Table 4.9. A full list of current meta-events can be obtained from the MMA. It is allowable for a manufacturer to include meta-events specific to a particular software package in a MIDI file, although this is only recommended if the standard MIDI file is to be used as the *normal* storage format by the software. In such a case the 'type' identifier should be set to &7F. A software package should expect to encounter events in MIDI files that it cannot deal with, and be able simply to ignore them, since either new event types may be defined after a package has been written, or a particular feature may be unimplemented.

A standardised meta-event format for lyrics has been published by the MMA as Recommended Practice RP-017, avoiding the confusion that used to be widespread regarding the

Table 4.9 A selection of common meta-event type identifiers

Type (hex)	Length	Description
00	02	Sequence number
01	Var	Text event
02	Var	Copyright notice text
03	Var	Sequence or track name
04	Var	Instrument name
05	Var	Lyric text (normally one syllable per event)
06	Var	Marker text (rehearsal letters, etc.)
07	Var	Cue point text
20	01	MIDI channel prefix (ties subsequent events to a particular channel, until the next channel event, MIDI or meta)
2F	00	End of track. (No data follows)
51	03	Set tempo (μs per quarter note)
54	05	Timecode location (hh:mm:ss:ff:100ths) of track start (following the MTC convention for hours)
58	04	Time signature (see below)
59	02	Key signature. First data byte denotes number of sharps (+ve value) or flats (−ve value). Second data byte denotes major (0) or minor (1) key
7F	Var	Sequencer-specific meta-event (see above)

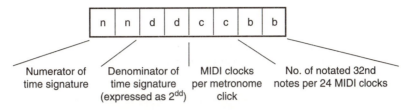

Figure 4.17 Meaning of the data bytes in the time signature meta-event

way that lyrics were represented in MIDI files. There is also a recommended practice for including device names and program names in meta-events, called RP-019. This enables specific destination devices to be identified in MIDI files, so that a track's events can subsequently be routed to that particular device. This is an alternative to identifying cable numbers on multi-port MIDI interfaces. The program name meta-event allows specific program or voice names to be included in the file so that the rather anonymous bank select and program change messages that are used to select sound generator voices can be identified by name.

4.10.5 Time signatures and tempo maps

The format of time signature meta-events needs further explanation, as it is somewhat arcane. The event consists of four data bytes following the 'length' identifier, as shown in Figure 4.17. The first two of these define the conventional time signature (e.g.: 4/4 or 6/8) and the second two define the relationship between MIDI clocks and the notated music. The denominator of the time signature is represented as the power of two required to produce the number concerned. For example, this value would be &03 if the denominator was 8_{10} because 2^3 equals 8. The third data byte defines the number of MIDI clocks per metronome click (the metronome may click at intervals other than a quarter note, depending on the time signature) and the final byte allows the user to define the number of 32nd notes actually notated per 24 MIDI clocks. This last, perhaps unusual-sounding definition allows for a redefinition of the tempo unit represented by MIDI clocks (which would normally run at a rate of 6 per 16th note), in order to accommodate software packages that allow this relationship to be altered. The tempo map of a song may need to be transferred between one machine and another, and the MIDI file format may be used for this purpose. Such a file could be a type 0 file consisting solely of meta-events describing tempo changes, but otherwise the map must be contained in the first track chunk of a larger file. This is where reading devices will expect to find it.

4.11 Downloadable Sounds (DLS) and SoundFonts

A gradual convergence may be observed in the industry between the various different methods by which synthetic sounds can be described. These have been variously termed 'Downloadable Sounds', 'SoundFonts' and more recently 'MPEG 4 Structured Audio Sample Bank Format'. Downloadable Sounds is an MMA specification for synthetic voice description that enables synthesisers to be programmed using voice data downloaded from a variety of sources. In this way a content creator could not only define the musical structure of his content in a universally usable way, using standard MIDI files, but could also define the nature of the sounds to be used with downloadable sounds. In these ways content creators can specify more precisely how synthetic audio should be replayed, so that the end result is more easily predicted across multiple rendering platforms.

The success of the most recent of these approaches depends to a large extent on the agreement around a common method of sound synthesis known as 'wavetable synthesis'. Here basic sound waveforms are stored in wavetables (simply tables of sample values) in RAM, to be read out at different rates and with different sample skip values, for replay at different pitches. Subsequent signal processing and envelope shaping can be used to alter the timbre and temporal characteristics. Such synthesis capabilities exist on the majority of computer sound cards, making it a realistic possibility to implement the standard widely.

Downloadable Sounds Level 1, version 1.1a, was published in 1999 and contains a specification for devices that can deal with DLS as well as a file format for containing the sound descriptions. The basic idea is that a minimal synthesis engine should be able to replay a looped sample from a wavetable, apply two basic envelopes for pitch and volume, use low frequency oscillator control for tremolo and vibrato, and respond to basic MIDI controls such as pitch bend and modulation wheel. There is no option to implement velocity crossfading or layering of sounds in DLS Level 1, but keyboard splitting into 16 ranges is possible.

DLS Level 1 requires a minimum specification of the sound card or rendering device that can be used to replay or render the synthetic sounds described. This includes a minimum of the following: 512 KB of wavetable storage (assuming 16 bit samples); 128 instruments, sets of articulation data, regions and samples at once; 24 simultaneous voices and 22.05 kHz sampling rate. The DLS Level 1 file format is based on the RIFF structure, described in Chapter 6. It is based on chunks containing instrument definitions and WAVE file data (containing the sampled audio for the individual wavetables). So-called articulation information describes things like loop points and envelope shapes that indicate how the sound is to be replayed. The WAVE data is mono and stored either in 16-bit twos complement form, or in 8-bit unsigned form. DLS RIFF files (file extension '.dls') may contain wave and articulation data for a number of instruments and for a number of note regions within those instruments. Information chunks provide textual information describing the instruments defined in the file, and instrument list chunks and sub-chunks contain the data for the instruments themselves – pointing to relevant wavetable data stored in WAVE chunks.

DLS Level 2 is somewhat more advanced, requiring two six-segment envelope generators, two LFOs, a low-pass filter with resonance and dynamic cut off frequency controls. It requires more memory for wavetable storage (2 MB), 256 instruments and 1024 regions, among other things. DLS Level 2 has been adopted as the MPEG-4 Structured Audio sample bank format (see Chapter 2 for more information about MPEG).

Emu developed so-called SoundFonts for Creative Labs and these have many similar characteristics to downloadable sounds. They have been used widely to define synthetic voices for Sound Blaster and other computer sound cards. In fact the formats have just about been harmonised with the issue of DLS Level 2 that apparently contains many of the advanced features of SoundFonts. SoundFont 2 descriptions are normally stored in RIFF files with the extension '.sf2'.

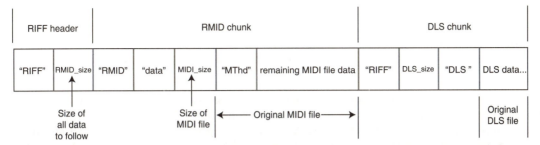

Figure 4.18 Structure of a basic RMID file containing both standard MIDI file (SMF) and downloadable sound (DLS) data. All header elements are 4 bytes long. The MIDI file data and the downloadable sound data are however long they are described to be

4.12 RMID and XMF files

RMID is a version of the RIFF file structure that can be used to combine a standard MIDI file and a downloadable sound file within a single structure. In this way all of the data required to replay a song using synthetic sounds can be contained within one file. As shown in Figure 4.18, it adds a 20-byte header to the file before the start of the SMF data, which contains the standard 4-byte ASCII 'RIFF' identification, followed by a 4-byte length indication and a 4-byte 'RMID' identifier. The 'data' chunk that follows contains the SMF data. A DLS chunk can be appended to the end of the SMF chunk within the overall RMID chunk length.

RMID seems to have been superseded by another file format known as XMF (eXtensible Music Format) that is designed to contain all of the assets required to replay a music file. It is based on Beatnik's RMF (Rich Music Format) which was designed to incorporate standard MIDI files and audio files such as MP3 and WAVE so that a degree of interactivity could be added to audio replay. RMF can also address a Special Bank of MIDI sounds (an extension of GM) in the Beatnik Audio Engine. XMF is now the MMA's recommended way of combining such elements. It is more extensible than RMID and can contain WAVE files and other media elements for streamed or interactive presentations. XMF introduces concepts such as looping and branching into standard MIDI files. RMF included looping but did not incorporate DLS into the file format. In addition to the features just described, XMF can incorporate 40-bit encryption for advanced data security as well as being able to compress standard MIDI files by up to 5:1 and incorporate metadata such as rights information. So far, XMF Type 0 and Type 1 have been defined, both of which contain SMF and DLS data, and which are identical except that Type 0 MIDI data may be streamed.

4.13 SAOL and SASL in MPEG 4 Structured Audio

SAOL is the Structured Audio Orchestra Language of MPEG 4 Structured Audio (a standard for low bit rate representation of digital audio). SASL is the Structured Audio Score Language. A SASL 'score' controls SAOL 'instruments'. SAOL is an extension of CSound, a synthesis language developed over many years, primarily at MIT, and is more advanced than MIDI DLS (which is based only on simple wavetable synthesis). Although there is a restricted

117

profile of Structured Audio that uses only wavetable synthesis (essentially DLS Level 2 for use in devices with limited processing power), a full implementation allows for a variety of other synthesis types such as FM, and is extensible to include new 'unit generators' (the CSound name for the elements of a synthesis patch).

SASL is more versatile than standard MIDI files in its control of SAOL instruments. There is a set of so-called 'MIDI semantics' that enables the translation of MIDI commands and controllers into SAOL events, so that MIDI commands can either be used instead of a SASL score, or in addition to it. If MPEG 4 Structured Audio (SA) gains greater ground and authoring tools become more widely available, the use of MIDI control and DLS may decline as they are inherently less versatile. MIDI, however, is inherently simpler than SA and could well continue to be used widely when the advanced features of SA are not required.

4.14 MIDI and synchronisation

4.14.1 Introduction to MIDI synchronisation

An important aspect of MIDI control is the handling of timing and synchronisation data. MIDI timing data takes the place of the various older standards for synchronisation on drum machines and sequencers that used separate 'sync' connections carrying a clock signal at one of a number of rates, usually described in pulses-per-quarter-note (ppqn). There used to be a considerable market for devices to convert clock signals from one rate to another, so that one manufacturer's drum machine could lock to another's sequencer, but MIDI has supplanted these by specifying standard synchronisation data that shares the same data stream as note and control information.

Not all devices in a MIDI system will need access to timing information – it depends on the function fulfilled by each device. A sequencer, for example, will need some speed reference to control the rate at which recorded information is replayed and this speed reference could either be internal to the computer or provided by an external device. On the other hand, a normal synthesiser, effects unit or sampler is not normally concerned with timing information, because it has no functions affected by a timing clock. Such devices do not normally store rhythm patterns, although there are some keyboards with onboard sequencers that ought to recognise timing data.

As MIDI equipment has become more integrated with audio and video systems the need has arisen to incorporate timecode handling into the standard and into software. This has allowed sequencers either to operate relative to musical time (e.g. bars and beats) or to 'real' time (e.g. minutes and seconds). Using timecode, MIDI applications can be run in sync with the replay of an external audio or video machine, in order that the long-term speed relationship between the MIDI replay and the machine remains constant. Also relevant to the systems integrator is the MIDI Machine Control standard that specifies a protocol for the remote control of devices such as external recorders using a MIDI interface.

4.14.2 Music-related timing data

This section describes the group of MIDI messages that deals with 'music-related' synchronisation – that is synchronisation related to the passing of bars and beats as opposed to 'real' time in hours, minutes and seconds. It is normally possible to choose which type of sync data will be used by a software package or other MIDI receiver when it is set to 'external sync' mode.

A group of system messages called the 'system realtime' messages control the execution of timed sequences in a MIDI system and these are often used in conjunction with the *song pointer* (which is really a system common message) to control autolocation within a stored song. The system realtime messages concerned with synchronisation, all of which are single bytes, are:

```
&F8 Timing clock

&FA Start

&FB Continue

&FC Stop
```

The timing clock (often referred to as 'MIDI beat clock') is a single status byte (&F8) to be issued by the controlling device six times per MIDI beat. A MIDI beat is equivalent to a musical semiquaver or sixteenth note (see Table 4.10) so the increment of time represented by a MIDI clock byte is related to the duration of a particular *musical* value, not directly to a unit of real time. Twenty-four MIDI clocks are therefore transmitted per quarter note, unless the definition is changed. (As mentioned in the discussion of time signature format in MIDI files (see Section 4.10.5) some software packages allow the user to redefine the notated musical increment represented by MIDI clocks.) At any one musical tempo, a MIDI beat could be said to represent a fixed increment of time, but this time increment would change if the tempo changed.

The timing clock byte, like other system realtime messages, may temporarily interrupt other MIDI messages, the status reverting to the previous status automatically after the realtime message has been handled by a receiver. This is necessary because of the very nature of the timing clock as a synchronising message. If it were made to wait for other messages to finish, it would lose its ability to represent a true increment of time. It may be seen that there could still be a small amount of error in the timing of any clock byte within the data stream if a large amount of other data was present, because the timing byte may not interrupt until at least the break between one byte and another, but this timing error cannot be greater than plus or minus half the duration of a MIDI byte, which is 160 µs.

Table 4.10 Musical durations related to MIDI timing data

Note value	Number of MIDI beats	Number of MIDI clocks
Semibreve (whole note)	16	96
Minim (half note)	8	48
Crotchet (quarter note)	4	24
Quaver (eighth note)	2	12
Semiquaver (sixteenth note)	1	6

So the &F8 byte might appear between the two data bytes of a note on message, for example, but it would not be necessary to repeat either the entire message or the 'note on' status after &F8 had passed. &F8 may also interrupt running status in the same way, without the need for reiteration of the status after the timing byte has been received. MIDI clocks should be given a very high priority by receiving software, since the degree of latency in the handling of this data will affect the timing stability of synchronised replay. On receipt of &F8, a device that handles timing information should increment its internal clock by the relevant amount. This in turn will increment the internal song pointer after six MIDI clocks (i.e. one MIDI beat) have passed. Any device controlling the sequencing of other instruments should generate clock bytes at the appropriate intervals and any changes of tempo within the system should be reflected in a change in the rate of MIDI clocks. In systems where continuously varying changes have been made in the tempo, perhaps to imitate *rubato* effects or to add 'human feel' to the music, the rate of the clock bytes will reflect this.

The 'start', 'stop' and 'continue' messages are used to remotely control the receiver's replay. A receiver should only begin to increment its internal clock or song pointer after it receives a start or continue message, even though some devices may continue to transmit MIDI clock bytes in the intervening periods. For example, a sequencer may be controlling a number of keyboards, but it may also be linked to a drum machine that is playing back an internally stored sequence. The two need to be locked together, so the sequencer (running in internal sync mode) would send the drum machine (running in external sync mode) a 'start' message at the beginning of the song, followed by MIDI clocks at the correct intervals thereafter to keep the timing between the two devices correctly related. If the sequencer was stopped it would send 'stop' to the drum machine, whereafter 'continue' would carry on playing from the stopped position, and 'start' would restart at the beginning. This method of synchronisation appears to be fairly basic, as it allows only for two options: playing the song from the beginning or playing it from where it has been stopped.

SPPs are used when one device needs to tell another where it is in a song. (The term 'song' is used widely in MIDI parlance to refer to any stored sequence.) A sequencer or synchroniser should be able to transmit song pointers to other synchronisable devices when a new location is required or detected. For example, one might 'fast-forward' through a song and start again twenty bars later, in which case the other timed devices in the system would have to know where to restart. An SPP would be sent followed by 'continue' and then regular clocks. Originally it was recommended that a gap of at least 5 seconds was left between sending a SPP and restarting the sequence, in order to give the receiver time to locate to the new position, but revisions state that a receiver should be able to register a 'continue' message and count subsequent MIDI clocks even while still locating, even if it is not possible to start playing immediately. Replay should begin as soon as possible, taking into account the clocks elapsed since the 'continue' message was received.

An SPP represents the position in a stored song in terms of number of MIDI beats (not clocks) from the start of the song. It uses two data bytes so can specify up to 16 384 MIDI beats. SPP is a system common message, not a realtime message. It is often used in conjunction with &F3 (song select), which is used to define which of a collection of stored song sequences (in a drum machine, say) is to be replayed. SPPs are fine for directing the movements of an entirely musical system, in which every action is related to a particular beat or subdivision of a beat, but not so fine when actions must occur at a particular point in real time. If, for example, one was using a MIDI system to dub music and effects to a picture in which an effect was intended to occur at a particular visual event, that effect would have to maintain its position in time no matter what happened to the music. If the effect was to be triggered by a sequencer at a particular number of beats from the beginning of the song, this point could change in real time if the tempo of the music was altered slightly to fit a particular visual scene. Clearly some means of real-time synchronisation is required either instead of, or as well as the clock and song pointer arrangement, such that certain events in a MIDI controlled system may be triggered at specific *times* in hours, minutes and seconds.

Recent software may recognise and be able to generate the bar marker and time signature messages. The bar marker message can be used where it is necessary to indicate the point at which the next musical bar begins. It takes effect at the next &F8 clock. Some MIDI synchronisers will also accept an audio input or a tap switch input so that the user can program a tempo track for a sequencer based on the rate of a drum beat or a rate tapped in using a switch. This can be very useful in synchronising MIDI sequences to recorded music, or fitting music which has been recorded 'rubato' to bar intervals.

4.14.3 Timecode and synchronisation

There are a number of ways of organising real-time synchronisation in a workstation, but they all depend on the use of timecode in one form or another. In this section the principles of timecode and its relationship to MIDI are explained.

Timecode is more correctly referred to as SMPTE/EBU time and control code. It is often just referred to as SMPTE ('simpty') in studios. It comes in two forms: linear timecode (LTC), which is an audio signal capable of being recorded on a tape recorder, and vertical interval timecode (VITC), which is recorded in the vertical interval of a television picture. Timecode is basically a binary data signal registering time from an arbitrary start point (which may be the time of day) in hours, minutes, seconds and frames, against which the program runs. It was originally designed for video editing, and every single frame on a particular video tape has its own unique number called the timecode address. This can be used to pinpoint a precise editing position. More recently timecode has found its way into audio, where TV frames have less meaning but are still used as a convenient subdivision of a second. Sometimes a sample offset is added to a timecode value to indicate the precise point of an edit in audio samples from the start of a frame.

A number of frame rates are available, depending on the television standard to which they relate, the frame rate being the number of still frames per second used to give the impression of continuous motion in the TV picture. Thirty frames per second (fps), or true SMPTE, was used for monochrome American television; 29.97 fps is used for colour NTSC television (mainly USA, Japan and parts of the Middle East), and is called 'SMPTE drop-frame'; 25 fps is used for PAL and SECAM TV and is called 'EBU' (Europe, Australia, etc.); and 24 fps is used for some film work. SMPTE drop frame timecode is so called because in order to maintain sync with NTSC colour television pictures running at 29.97 fps it is necessary to use the 30 fps SMPTE code but to drop two frames at the start of each minute, except every tenth minute. This is a compromise solution which has the effect of introducing a short term sync error between timecode and real time, whilst maintaining reasonable control over the long-term drift.

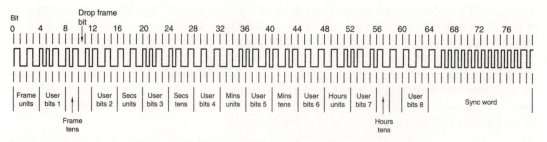

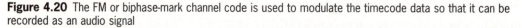

Figure 4.19 Data format of the SMPTE/EBU longitudinal timecode frame. Note the sync word 0011111111111101 which occurs at the end of each frame to mark the boundary. This pattern does not occur elsewhere in the frame and its asymmetry allows a timecode reader to determine the direction in which the code is being played

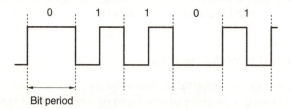

Figure 4.20 The FM or biphase-mark channel code is used to modulate the timecode data so that it can be recorded as an audio signal

An LTC frame value is represented by an 80-bit binary 'word', split principally into groups of 4 bits, with each 4 bits representing a particular parameter such as tens of hours, units of hours, and so forth, in BCD (binary-coded decimal) form (see Figure 4.19). Sometimes, not all four bits per group are required – the hours only go up to '23', for example – and in these cases the remaining bits are either used for special control purposes or set to zero (unassigned): 26 bits in total are used for time address information to give each frame its unique hours, minutes, seconds, frame value; 32 are 'user bits' and can be used for encoding information such as reel number, scene number, day of the month and the like; bit 10 denotes drop-frame mode if a binary 1 is encoded there, and bit 11 can denote colour frame mode if a binary 1 is encoded (used in video editing). The end of each word consists of 16 bits in a unique sequence, called the 'sync word', and this is used to mark the boundary between one frame and the next. It also allows a timecode reader to tell in which direction the code is being read, since the sync word begins with '11' in one direction and '10' in the other.

This binary information cannot be directly recorded as an audio signal, since its bandwidth would be too wide, so it is modulated in a simple scheme known as 'biphase mark', or FM, in which a transition from one state to the other (low to high or high to low) occurs at the edge of each bit period, but an additional transition is forced within the period to denote a binary 1 (see Figure 4.20). The result looks rather like a square wave with two frequencies, depending on the presence of ones and zeros in the code. The code can be read forwards or backwards, and phase inverted. Readers are available which will read timecode over a very wide range of speeds, from around 0.1 to 200 times play speed. The rise-time of the signal, that is the time it takes to swing between its two extremes, is specified as 25 μs ± 5 μs, and this requires an audio bandwidth of about 10 kHz.

VITC is recorded not on an audio track, but in the vertical sync period of a video picture, such that it can always be read when video is capable of being read, such as in slow-motion and pause modes. It is useful in applications where slow-motion cueing is to be used in the location of sync or edit points and is extracted directly from the video signal by a timecode reader. Some MIDI synchronisers can accept VITC, but this is much less common than the ability to read and write LTC.

In audio workstations timecode is not usually recorded as an audio signal on a specific 'track' but is derived from the system clock in relation to the replay rate of an audio or MIDI sequence. Its use as an audio signal (LTC) will probably decline as more and more synchronisation of audio, video and MIDI takes place within the workstation itself. LTC will remain useful for synchronisation with external recorders.

4.14.4 MIDI timecode (MTC)

MIDI timecode has two specific functions. Firstly, to provide a means for distributing conventional SMPTE/EBU timecode data around a MIDI system in a format that is compatible with the MIDI protocol. Secondly, to provide a means for transmitting 'setup' messages that may be downloaded from a controlling computer to receivers in order to program them with cue points at which certain events are to take place. The intention is that receivers will then read incoming MTC as the program proceeds, executing the pre-programmed events defined in the setup messages. Sequencers and some digital audio systems often use MIDI timecode derived from an external synchroniser or MIDI peripheral when locking to video or to another sequencer. MTC is an alternative to MIDI clocks and song pointers, for use when real time synchronisation is important.

In an LTC timecode frame, two binary data groups are allocated to each of hours, minutes, seconds and frames, these groups representing the tens and units of each, so there are eight binary groups in total representing the time value of a frame. In order to transmit this information over MIDI, it has to be turned into a format that is compatible with other MIDI data (i.e. a status byte followed by relevant data bytes). There are two types of MTC synchronising message: one that updates a receiver regularly with running timecode and another that transmits one-time updates of the timecode position. The latter can be used during high speed cueing, where regular updating of each single frame would involve too great a rate of transmitted data. The former is known as a quarter-frame message, denoted by the status byte (&F1), whilst the latter is known as a full-frame message and is transmitted as a universal realtime SysEx message.

One timecode frame is represented by too much information to be sent in one standard MIDI message, so it is broken down into eight separate messages. Each message of the group of eight represents a part of the timecode frame value, as shown in Figure 4.21, and takes the general form:

```
&[F1]  [DATA]
```

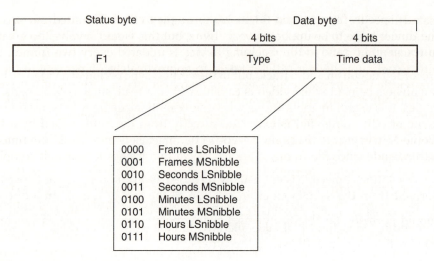

Figure 4.21 General format of the quarter-frame MTC message

The data byte begins with zero (as always), and the next seven bits of the data word are made up of a 3-bit code defining whether the message represents hours, minutes, seconds or frames, MSnibble or LSnibble, followed by the four bits representing the binary value of that nibble. In order to reassemble the correct timecode value from the eight quarter-frame messages, the LS and MS nibbles of hours, minutes, seconds and frames are each paired within the receiver to form 8-bit words as follows:

```
Frames: rrr qqqqq
```

where 'rrr' is reserved for future use and 'qqqqq' represents the frames value from 0 to 29;

```
Seconds: rr qqqqqq
```

where 'rr' is reserved for future use and 'qqqqqq' represents the seconds value from 0 to 59;

```
Minutes: rr qqqqqq
```

as for seconds; and

```
Hours: r qq ppppp
```

where 'r' is undefined, 'qq' represents the timecode type, and 'ppppp' is the hours value from 0 to 23. The timecode frame rate is denoted as follows in the 'qq' part of the hours value: 00 = 24 fps; 01 = 25 fps; 10 = 30 fps drop-frame; 11 = 30 fps non-drop-frame. Unassigned bits should be set to zero.

At a frame rate of 30 fps, quarter-frame messages would be sent over MIDI at a rate of 120 messages per second. As eight messages are needed fully to represent a frame, it can be appreciated that $30 \times 8 = 240$ messages really ought to be transmitted per second if the receiving device were to be updated every frame, but this would involve too great an overhead in transmitted data, so the receiving device is updated every two frames. If MTC is transmitted continuously over MIDI it takes up approximately 7.5 per cent of the available data bandwidth. Quarter-frame messages can be transmitted in forward or reverse order, to emulate timecode running either forwards or backwards, with the 'frames LSnibble' message transmitted on the frame boundary of the timecode frame that it represents.

The receiver must in fact maintain a two-frame offset between displayed timecode and received timecode since the frame value has taken two frames to transmit completely. For real-time synchronisation purposes, the receiver may wish simply to note that time has advanced another quarter of a frame at the receipt of each quarter-frame message, rather as it advances by one-sixth of a beat on receipt of each MIDI clock. Internal synchronisation software should normally be able to flywheel or interpolate between received synchronisation messages in order to obtain higher internal resolution than that implied by the rate of the messages. For all except the fastest musical tempo values, MIDI timecode messages arrive more regularly than MIDI clocks would, so they might be considered a more reliable timing reference. Nonetheless, MIDI clocks are still needed when synchronisation is based on musical time increments.

The format of the full-frame message is as follows, falling into the group of messages known as the sysex universal realtime messages:

```
&[F0] [7F] [dev. ID] [01] [01] [hh] [mm] [ss] [fr] [F7]
```

The device ID would normally be set to &7F which signifies that the message is intended for the whole system, the sub-ID #1 of &01 denotes an MTC message, and sub-ID #2 denotes a full-frame message. Thereafter hours, minutes, seconds and frames take the same form as for quarter-frame messages.

4.15 MIDI machine control (MMC)

MIDI may be used for remotely controlling tape machines and other studio equipment, as well as musical instruments. MMC uses universal realtime SysEx messages with a sub-ID #1 of either &06 or &07 and has a lot in common with a remote control protocol known as 'ESbus' which was devised by the EBU and SMPTE as a universal standard for the remote control of tape machines, VTRs and other studio equipment. The ESbus standard uses an RS422 remote control bus running at 38.4 kbaud, whereas the MMC standard uses the MIDI bus for similar commands. Although MMC and ESbus are not the same and the message protocols are not identical, the command types and reporting capabilities required of machines are very similar. There are a number of levels of complexity at which MMC can be made to operate, making it possible for people to implement it at anything from a very simple level (i.e. cheaply) to a very complicated level involving all the finer points.

MMC is designed to work in either open- or closed-loop modes (see Figure 4.22). This is similar to other system exclusive applications that make use of handshaking between the transmitter and the receiver. Communication can be considered as occurring between a

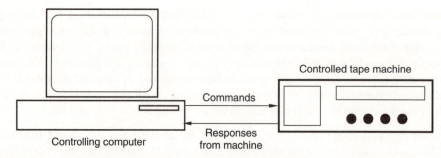

Controlled tape machine

Commands

Controlling computer

Responses
from machine

Figure 4.22 A closed-loop MMC arrangement. The controller should receive a response from the controlled device within two seconds of issuing a command which expects a response. If it does not, it should assume an open loop

'controller' and a 'controlled device', with commands flowing from the controller to the controlled device and responses returning in the opposite direction. Since a controller may address more than one controlled device at a time it is possible for a number of responses to be returned, and this situation requires careful handling, as discussed below. It is expected that MMC devices and applications will default to the closed-loop condition, but a controller should be able to detect an open-loop situation by timing out if it does not receive a response within two seconds after it has sent a message which requires one. From then on, an open loop should be assumed. Alternatively, a controller could continue to check for the completion of a closed loop by sending out regular requests for a response, changing modes after a response was received. In the closed-loop mode a simple handshaking protocol is defined, again similar in concept to the sample and file dump modes, but involving only two messages – WAIT and RESUME. These handshaking messages are used to control the flow of data between controller and controlled device in both directions, in order to prevent the overflowing of MIDI receive buffers (which would result in loss of data). Handshaking is discussed further below.

Typical MMC communications involve the transmission of a command from the controller to a particular device, using its device ID as a means of identifying the destination of the command. It is also possible to address all controlled devices on the bus using the &7F device ID in place of the individual ID. Commands take the general format:

```
&[F0]  [7F]  [dev. ID]  [06]  [data]  ...  ...  [F7]
```

Note that only sub-ID #1 is used here, following the device ID, and there is no sub-ID #2 in order to conserve data bandwidth. The sub-ID #1 of &06 denotes an MMC command. [data] represents the data messages forming the command, and may be from one to many bytes in length.

The amount of data making up a command depends on its type. Commands that consist of only a single byte, such as the 'play' command (&02), occupy the range from &01 to &3F (&00 is reserved to be used for future extensions to the command set). A typical command of this type (e.g. 'play') would thus be transmitted as:

```
&[F0]  [7F]  [dev. ID]  [06]  [02]  [7F]
```

The handshaking messages, WAIT (&7C) and RESUME (&7F), can be issued by either the controller or any of its controlled devices. Handshaking depends on the use of a closed loop. When issued by the controller the message would normally be a command addressed to any device trying to send data back to it, and thus the device ID attached to controller handshaking messages is &7F ('all call'). For example, a controller whose receive data buffer was approaching overflow would wish to send out a general 'everybody WAIT' command, to suspend MMC transmission from controlled devices until it had reduced the contents of the buffer, after which an 'everybody RESUME' command would be transmitted. Such a command would take the form:

```
&[F0] [7F] [7F] [06] [7C or 7F] [F7]
```

When issued by a controlled device, handshaking messages should be a response tagged with the device's own ID, as a means of indicating to the controller *which* device is requesting a WAIT or RESUME. On receipt of a WAIT from a particular device ID the controller would suspend transmissions to that device but continue to transmit commands to others. Such a message would take the form:

```
&[F0] [7F] [dev. ID] [07] [7C or 7F] [F7]
```

Table 4.11 gives a list of the single byte transport commands used in the MMC protocol. The list of MMC commands and their accompanying data occupies many pages so it is not proposed to describe them in detail here. Readers should refer to the MIDI Machine Control section of the MIDI standard for the latest information. There is no mandatory set of commands or responses defined in the standard, although there are some guidelines concerning possible minimum sets for certain applications. It is possible to tell which MMC commands and responses have been implemented in a particular device by analysing the 'signature' of the device. The signature will normally be both published in written form in the manual, and available as a response from the controlled device. It exists in the form of a bit map in which each bit corresponds to a certain MMC function. If the bit is set to '1' then the function is implemented. The signature comes in two parts: the first describing the commands implemented and the second describing the responses implemented. It also contains a header

Table 4.11 Basic MMC transport controls

Command	Hex value	Comment
Stop	01	
Play	02	
Deferred play	03	Play after autolocate achieved
Fast fwd	04	
Rewind	05	
Record strobe	06	Drop into or out of record (depending on rec. ready state)
Record exit	07	
Record pause	08	Enters record-pause mode
Pause	09	

describing the version of MMC used in the device. The exact format of the signature is described in the MMC standard.

4.16 MIDI over USB

USB (Universal Serial Bus) is a computer peripheral interface that carries data at a much faster rate than MIDI (up to 12 Mbit s^{-1} or up to 480 Mbit s^{-1}, depending on the version). It is very widely used on workstations and peripherals these days and it is logical to consider using it to transfer MIDI data between devices as well. The USB Implementers Forum has published a 'USB Device Class Definition for MIDI Devices', version 1.0, that describes how MIDI data may be handled in a USB context. It preserves the protocol of MIDI messages but packages them in such a way as to enable them to be transferred over USB. It also 'virtualises' the concept of MIDI IN and OUT jacks, enabling USB to MIDI conversion, and vice versa, to take place in software within a synthesiser or other device. Physical MIDI ports can also be created for external connections to conventional MIDI equipment (see Figure 4.23).

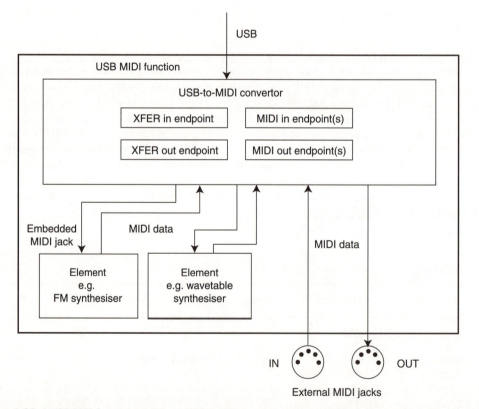

Figure 4.23 A USB MIDI function contains a USB-to-MIDI convertor that can communicate with both embedded (internal) and external MIDI jacks via MIDI IN and OUT endpoints. Embedded jacks connect to internal elements that may be synthesisers or other MIDI data processors. XFER in and out endpoints are used for bulk dumps such as DLS and can be dynamically connected with elements as required for transfers

USB packet header		Normal MIDI message		
Cable number	Code Index Number	MIDI_0	MIDI_1	MIDI_2

Figure 4.24 USB MIDI packets have a 1-byte header that contains a cable number to identify the MIDI jack destination and a code index number to identify the contents of the packet and the number of active bytes

A so-called 'USB MIDI function' (a device that receives USB MIDI events and transfers) may contain one or more 'elements'. These elements can be synthesisers, synchronisers, effects processors or other MIDI-controlled objects.

A USB to MIDI convertor within a device will typically have MIDI in and out endpoints as well as what are called 'transfer' (XFER) endpoints. The former are used for streaming MIDI events whereas the latter are used for bulk dumps of data such as those needed for downloadable sounds (DLS). MIDI messages are packaged into 32-bit USB MIDI events, which involve an additional byte at the head of a typical MIDI message. This additional byte contains a cable number address and a code index number (CIN), as shown in Figure 4.24. The cable number enables the MIDI message to be targeted at one of 16 possible 'cables', thereby overcoming the 16-channel limit of conventional MIDI messages, in a similar way to that used in the addressing of multiport MIDI interfaces. The CIN allows the type of MIDI message to be identified (e.g. System Exclusive; Note On), which to some extent duplicates the MIDI status byte. MIDI messages with fewer than three bytes should be padded with zeros.

The USB message transport protocol and interfacing requirements are not the topic of this book, so users are referred to the relevant USB standards for further information about implementation issues.

4.17 MIDI over IEEE 1394

IEEE 1394, or 'Firewire' as it is sometimes known, is another high speed serial interface encountered widely on workstations and media equipment. It is capable of data rates up to many hundreds of megabits per second. The MMA and AMEI have published a 'MIDI Media Adaptation Layer for IEEE 1394' that describes how MIDI data may be transferred over 1394. This is also referred to in 1394 TA (Trade Association) documents describing the 'Audio and Music Data Transmission Protocol' and IEC standard 61883-6 that deals with the audio part of 1394 interfaces.

The approach is similar to that used with USB, described Section 4.16, but has somewhat greater complexity. MIDI 1.0 data streams can be multiplexed into a 1394 'MIDI conformant data channel' that contains eight independent MIDI streams called 'MPX-MIDI data channels'. This way each MIDI conformant data channel can handle $8 \times 16 = 128$ MIDI channels (in the original sense of MIDI channels). The data are transferred in AM824 format (see Section 6.7.1), using groups of 'quadlets' (four bytes). The first version of the standard limits the transmission of packets to the MIDI 1.0 data rate of 31.25 kbit s^{-1} for compatibility with other MIDI devices, however provision is made for transmission at substantially faster rates for use in equipment that is capable of it. This includes options for 2X and 3X MIDI 1.0 speed.

1394 cluster events can be defined that contain both audio and MIDI data. This enables the two types of information to be kept together and synchronised.

4.18 After MIDI?

Various alternatives have been proposed over the years, aiming to improve upon MIDI's relatively limited specification and flexibility when compared with modern music control requirements and computer systems. That said, MIDI has shown surprising robustness to such 'challenges' and has been extended over the years so as to ameliorate some of its basic problems. Perhaps the simplicity and ubiquity of MIDI has made it attractive for developers to find ways of working with old technology that they know rather than experimenting with untried but more sophisticated alternatives.

ZIPI was a networked control approach proposed back in the early 1990s that aimed to break free from MIDI's limitations and take advantage of faster computer network technology, but it never really gained widespread favour in commercial equipment. It has now been overtaken by more recent developments and communication buses such as USB and 1394.

Open Sound Control is currently a promising alternative to MIDI that is gradually seeing greater adoption in the computer music and musical instrument control world. Developed by Matt Wright at CNMAT (Centre for New Music and Audio Technology) in Berkeley, California, it aims to offer a transport-independent message-based protocol for communication between computers, musical instruments and multimedia devices. It does not specify a particular hardware interface or network for the transport layer, but initial implementations have tended to use UDP (user datagram protocol) over Ethernet or other fast networks as a transport means. It is not proposed to describe this protocol in detail and further details can be found at the website indicated at the end of this chapter. A short summary will be given, however.

OSC uses a form of device addressing that is very similar to an Internet URL (uniform resource locator). In other words a text address with sub-addresses that relate to lower levels in the device hierarchy. For example (not a real address) '/synthesiser2/voice1/oscillator3/frequency' might refer to a particular device called 'synthesiser2', within which is contained voice 1, within which is oscillator 3, whose frequency value is being addressed. The minimum 'atomic unit' of OSC data is 4 bytes (32 bits) long, so all values are 32-bit aligned, and transmitted packets are made up of multiples of 32-bit information. Packets of OSC data contain either individual messages or so-called 'bundles'. Bundles contain elements that are either messages or further bundles, each having a size designation that precedes it, indicating the length of the element. Bundles have time tags associated with them, indicating that the actions described in the bundle are to take place at a specified time. Individual messages are supposed to be executed immediately. Devices are expected to have access to a representation of the correct current time so that bundle timing can be related to a clock.

Further reading

Hewlett, W. and Selfridge-Field, E. (eds) (2001) *The Virtual Score: Representation, Retrieval, Restoration.* MIT Press.

MMA (1999) *Downloadable Sounds Level 1. V1.1a*, January. MIDI Manufacturers Association.

MMA (2000) *RP-027: MIDI Media Adaptation Layer for IEEE 1394*. MIDI Manufacturers Association.

MMA (2000) *RP-029: Bundling SMF and DLS data in an RMID file*. MIDI Manufacturers Association.

MMA (2001) *XMF Specification Version 1.0*. MIDI Manufacturers Association.

MMA (2002) *The Complete MIDI 1.0 Detailed Specification*. MIDI Manufacturers Association.

MMA (2002) *Scalable polyphony MIDI specification and device profiles*. MIDI Manufacturers Association.

Scheirer, D. and Vercoe, B. (1999) SAOL: the MPEG 4 Structured Audio Orchestra Language. *Computer Music Journal*, **23**, 2, pp. 31–51.

Selfridge-Field, E., Byrd, D. and Bainbridge, D. (1997) *Beyond MIDI: The Handbook of Musical Codes*. MIT Press.

USB Implementers' Forum (1996) *USB Device Class Definition for MIDI Devices, version 1.0*. Available from www.usb.org.

Useful websites

MIDI Manufacturers Association: www.midi.org

Music XML: www.musicxml.org

Open Sound Contol: cnmat.cnmat.berkeley.edu/OpenSoundControl/

5 Hardware and systems issues

This chapter is concerned with explaining the audio storage media, sound cards and interfaces commonly encountered on computers. It also considers signal processing options and hardware abstraction layers for communication with sound cards.

5.1 Storage media

The purpose of this section is to describe the principles, limitations and applications of storage media used for audio in computer workstations. The media described are not exclusive to the field of audio and are widely encountered in general-purpose storage applications. In most cases the same media that are used for general purpose applications in computers can be used for storing audio and video without modification, although certain specifications must be adequate if operation is to be satisfactory. There will continue to be a decline in the use of dedicated audio recording formats in favour of general purpose mass storage media, if only because of the simple economics of the matter.

Improvements in the design of storage media will continue and prices will continue to fall. The devices described here are likely to remain popular for some years to come and in any case the fundamental principles involved are unlikely to change radically. Examples of specifications should only be taken as representative of today's equipment.

5.1.1 Storage requirements of digital audio and video

There are two main roles for storage media in audio workstations. One is the primary role of real-time recording and replay and the other is the secondary role of backup storage. The requirements differ somewhat, although it is possible to use similar media for both purposes. Real-time recording and replay needs storage devices capable of sustaining data transfer for a number of audio channels, so that the channels can record or replay for long periods without breaks, be edited and post processed, with quick access to stored files. This was discussed in greater detail in Chapter 3. Backup can take place in non-real time, does not need such fast access to files and does not need to support editing and other post processing operations. Backup may also need a large capacity and it would be advantageous if it were cheaper than primary storage, and be based on removable media. It follows that certain devices are suitable for backup that may not be suitable for primary storage.

Storage systems may use removable media but many have fixed media. It is advantageous to have removable media for audio and video purposes because it allows different jobs to be kept on different media and exchanged at will, but unfortunately the highest performance is normally only obtainable from storage systems with fixed media. Systems involving a small number of audio channels or using data reduction may be able to take advantage of removable media as primary storage, but in most current systems removable media are normally used as secondary storage.

It perhaps goes without saying that any storage system used for audio and video should be as reliable and robust as possible. It is also likely to need to be a fairly 'heavy duty' system because the demands of audio and video recording are quite heavy and will require the storage device to be in an almost constant state of activity. This differs from the more gentle task of, say, word processing, where the storage device is idling for long periods.

5.1.2 Disk drives in general

Disk drives are probably the most common form of mass storage. They have the advantage of being random-access systems – in other words any data can be accessed at random and with only a short delay. This may be contrasted with tape drives that only allow linear access – by winding through the tape until the desired data is reached, resulting in a considerable delay. Disk drives come in all shapes and sizes from the commonly encountered floppy disk at the bottom end to high performance hard drives at the top end. The means by which data are stored is usually either magnetic or optical, but some use a combination of the two, as described below. There exist both removable and fixed media disk drives, but in almost all cases the fixed media drives have a higher performance than removable media drives. This is because the design tolerances can be made much finer when the drive does not have to cope with removable media, allowing higher data storage densities to be achieved. Although removable disk media can appear to be expensive compared with tape media, the cost must be weighed against the benefits of random access and the possibility that some removable disks can be used for primary storage whereas a tape can not. Removable media should be distinguished from removable drives, the latter requiring that the complete drive is removed from the system as opposed to the storage surface(s) only.

The general structure of a disk drive is shown in Figure 5.1. It consists of a motor connected to a drive mechanism that causes one or more disk surfaces to rotate at anything from a few hundred to many thousands of revolutions per minute. This rotation may either remain constant or may stop and start, and it may either be at a constant rate or a variable rate, depending on the drive. One or more heads are mounted on a positioning mechanism that can move the head across the surface of the disk to access particular points, under the control of hardware and software called a disk controller. The heads read data from and write data

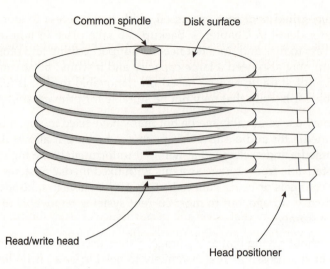

Figure 5.1 The general mechanical structure of a disk drive

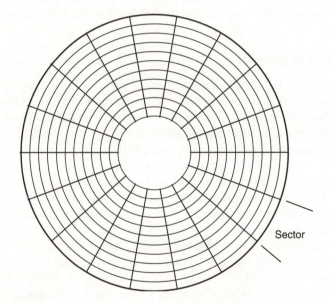

Figure 5.2 Disk formatting divides the storage area into tracks and sectors

to the disk surface by whatever means the drive employs. Certain disk types are read-only, some are write-once-read-many (WORM) and some are fully erasable and rewritable.

The disk surface is normally divided up into tracks and sectors, not physically but by means of 'soft' formatting (see figure 5.2). Formatting writes logical markers to indicate block boundaries, amongst other processes. On most hard disks the tracks are arranged as a series of concentric rings, but with some optical disks there is a continuous spiral track.

5.1.3 Disk drive specifications

Disk drive performance is characterised by specifications that are often quoted in promotional literature. These are the subject of a certain amount of misunderstanding and manufacturers often play games with these figures to make their drives seem better than they are. As with all specifications it is important to compare like with like, and to know how a certain parameter has been measured. The most important parameters are:

- access time;
- instantaneous transfer rate;
- sustained transfer rate; and
- storage capacity (formatted).

These are not the only factors that affect the performance or desirability of a drive, but they are a ready means of comparing two apparently similar drives.

Access time, normally quoted in milliseconds, is the time taken for a block of data to be accessed. It may be specified in a number of ways, since clearly the actual access time depends on where the head is when a block is requested. Figure 5.3 shows that true access time is made up of seek latency and rotational latency. The seek latency is dependent on the speed of the positioner and the rotational latency is dependent on how fast the disk rotates. Access time may often be just seek latency and may be quoted as 'track-to-track', which is the fastest, 'average', which is a reliable guide to general performance, or 'one-third full sweep', which is the time taken for the head to traverse one third of the active disk radius.

Instantaneous transfer rate is the fastest speed at which data can be read from the disk surface once the head has arrived at its correct location. Normally quoted in megabits per second, it gives a guide to the peak performance of the drive.

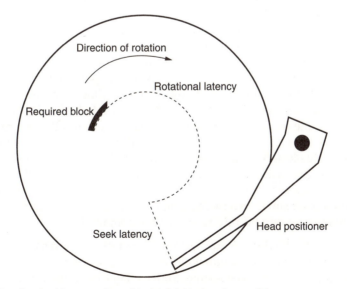

Figure 5.3 The delays involved in accessing a block of data stored on a disk

Sustained transfer rate is a more useful guide to real performance, though, because it gives a guide to the long-term data rate that might be expected from the disk, sustained over many blocks. This parameter, though, is affected considerably in real multimedia systems by the fragmentation of the drive and by the number of channels it has to service.

Formatted storage capacity is the number of megabytes of capacity available for user data after the disk has been formatted. It is often considerably smaller than the unformatted capacity of the disk (which is not a very useful figure to know). The formatted capacity is available for the storage of audio data if necessary, with no necessity to add an overhead for error correction, as described in Chapter 2.

5.1.4 Magnetic hard disk drives

Magnetic hard disks provide space for the storage of a large amount of data in a relatively small space, are reliable, fast and reasonably economical. Performance and capacity are normally in excess of typical multichannel audio requirements these days. One can store many hours of monophonic audio on a hard disk and they are capable of handling a large number of simultaneous channels of recording and replay. A quiet drive is important for audio operations, especially if the drive is to be installed in the same room as the operator.

A typical drive is a sealed unit and the physical disks inside it cannot be removed to make way for others. The recording process is magnetic, whereby data is stored in the form of flux reversals in the surface layer of the disks. The drive is a combination of physical disk surfaces on which data is stored, electromagnetic heads that read and write data, a positioner to move the heads to the right place, a motor that rotates the surfaces, a servo mechanism that controls the moving parts, and a controller that looks after the data flow to and from the surfaces and interfaces to the rest of the computer system. A cut-away diagram of an older drive is shown in Figure 5.4.

The drive is sealed (except sometimes for a small pressure-relief vent) in order to prevent the surfaces of the disks from becoming contaminated. The lack of contamination and the fact that the disks will never be removed means that fine tolerances can be used in manufacture, allowing a larger amount of data to be stored in a smaller space than is possible with removable magnetic disks. It also results in a very low error rate. One or more disks normally reside inside a drive and it is common for both sides of each disk to be used. These disks are rigid, not floppy, and all rotate on a common spindle. Each surface has its own read/write heads, which can be moved across the disk surface to access data stored in different places. The head positioner moves all the heads at the same time, rather than independently. The heads do not touch the surface of the disks during operation, they fly just a small distance above the surface, lifted by the aerodynamic effect of the air which is dragged around above the disk surface due to friction. A small area of the disk surface is set aside for the heads to land on when the power is turned off and this area does not contain data.

Data are stored in tracks divided up into sectors. Each sector is separated by a small gap and preceded by an address mark that uniquely identifies the sector's location and a preamble to synchronise the reading of data. The term cylinder relates to all the tracks that reside physically in line with each other in the vertical plane through the different surfaces (see Figure 5.5).

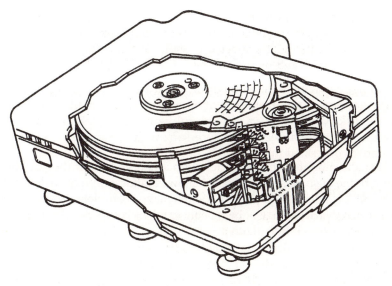

Figure 5.4 Cut-away drawing of a typical Winchester drive. (Courtesy of MacUser)

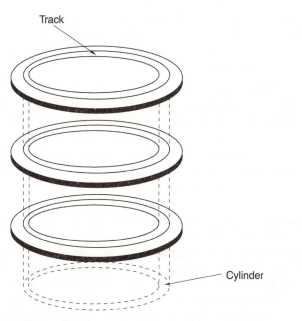

Track

Cylinder

Figure 5.5 Winchester drive tracks on different surfaces form concentric cylinders

A sector typically contains 512 bytes. The disk is of the 'write-many-times' format which means that old data may be overwritten many times in order to reuse the storage space. Although the disk surfaces of such a drive are not removable, drives exist that may be inter-changed in their entirety. Such drives are known as removable drives (not removable disks)

Figure 5.6 A typical removable disk drive system allowing multiple drives to be inserted or removed from the chassis at will (Courtesy of Glyph Technologies, Inc.)

and they are usually mounted in a cartridge with a handle so that they can be 'unplugged' from a docking frame of some sort. Figure 5.6 shows a photograph of such a system. This is a useful feature, but it is relatively expensive to interchange complete drives in this way. It may be considered worth the advantage of being able to take a complete session's primary storage from one system and insert it into another.

5.1.5 RAID arrays

Hard disk drives can be combined in various ways to improve either data integrity or data throughput. RAID stands for redundant array of inexpensive disks, and is a means of linking ordinary disk drives under one controller so that they form an array of data storage space, as shown in Figure 5.7. A RAID array can be treated as a single volume by a host computer. There are a number of levels of RAID array, each of which is designed for a slightly different purpose, as summarised in Table 5.1.

One of the main reasons for using a RAID array would be to improve the reliability of data storage. At certain RAID levels the data is spread across all of the drives involved, with a final drive used to store error protection information (the check drive). The aim is to prevent you losing your data if one of the drives fails, because it can be reconstructed from the remaining data. 'Mirroring' is also an option that allows the data on one disk to be perfectly duplicated on another, again for improving data security. By spreading data across drives it is also possible to speed up read and write operations.

5.1.6 Removable magnetic media

Floppy disks are unsuitable for AV applications because of just about every aspect of their specification. They are too small and too slow. Higher capacity removable magnetic media have existed for some time, though, with speeds approaching that of slower hard disks. These include things like Iomega's Zip disks that are constructed rather like large floppy disks, in a rigid cartridge. These have tended to offer capacities up to 250 Mbytes, which makes them only marginally useful for AV applications requiring short storage times.

Advances in the magnetic recording field have resulted in removable media offering much higher capacities and transfer speeds. Because removable media are not permanently sealed the reliability and performance may be less satisfactory than sealed hard disks, but there is the advantage of removability. One example, the Iomega Jaz drive, has performance suitable for primary storage in audio systems. This is a form of removable cartridge that houses hard disk

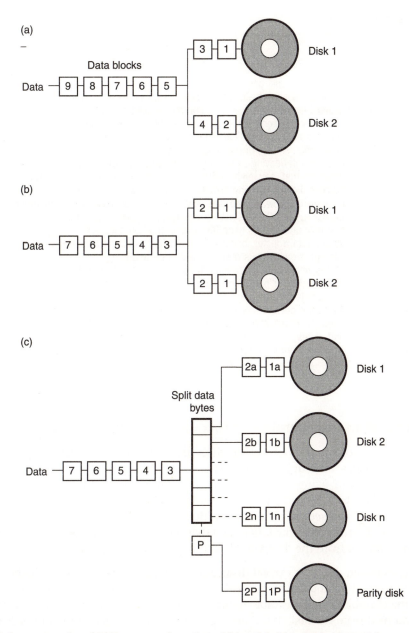

Figure 5.7 Some examples of RAID array configurations. (a) Level 0. (b) Level 1. (c) Level 3

platters inside a dustproof case. Syquest has also manufactured a range of high capacity removable storage systems (but they were bought out by Iomega), and Castlewood Systems has introduced a range of so-called 'ORB' drives based on magneto-resistive head technology that allows greater capacity per area of the disk surface.

Table 5.1 RAID levels

RAID level	Features
0	Data blocks split alternately between a pair of disks, but no redundancy so actually less reliable than a single disk. Transfer rate is higher than a single disk. Can improve access times by intelligent controller positioning of heads so that next block is ready more quickly
1	Offers disk mirroring. Data from one disk is automatically duplicated on another. A form of real-time backup
2	Uses bit interleaving to spread the bits of each data word across the disks, so that, say, eight disks each hold one bit of each word, with additional disks carrying error protection data. Non-synchronous head positioning. Slow to read data, and designed for mainframe computers
3	Similar to level 2, but synchronises heads on all drives, and ensures that only one drive is used for error protection data. Allows high speed data transfer, because of multiple disks in parallel. Cannot perform simultaneous read and write operations
4	Writes whole blocks sequentially to each drive in turn, using one dedicated error protection drive. Allows multiple read operations but only single write operations
5	As level 4 but splits error protection between drives, avoiding the need for a dedicated check drive. Allows multiple simultaneous reads and writes
6	As level 5 but incorporates RAM caches for higher performance

5.1.7 Optical disks in general

There are a number of families of optical disk drive that have differing operational and technical characteristics, although they share the universal benefit of removable media. They are all written and read using a laser, which is a highly focused beam of coherent light, although the method by which the data is actually stored varies from type to type. Optical disks are sometimes enclosed in a plastic cartridge that protects the disk from damage, dust and fingerprints, and they have the advantage that the pickup never touches the disk surface making them immune from the 'head crashes' that can affect magnetic hard disks.

Compatibility between different optical disks and drives is something of a minefield because the method of formatting and the read/write mechanism may differ. The most obvious differences lie in the erasable or non-erasable nature of the disks and the method by which data is written to and read from the disk, but there are also physical sizes and the presence or lack of a cartridge to consider. Drives tend to split into two distinct families from a compatibility point of view: those that handle CD/DVD formats and those that handle magneto-optical (M-O) and other cartridge-type ISO standard disk formats. The latter may be considered more suitable for 'professional purposes' whereas the former are often encountered in consumer equipment.

WORM disks (for example the cartridges that were used quite widely for archiving in the late 1980s and 90s) may only be written once by the user, after which the recording is permanent (a CD-R is therefore a type of WORM disk). Other types of optical disks can be written numerous times, either requiring pre-erasure or using direct overwrite methods (where new data is simply written on top of old, erasing it in the process). The read/write process of most current rewritable disks is typically 'phase change' or 'magneto-optical'. The CD-RW is an example of a rewritable disk that now uses direct overwrite principles.

The speed of some optical drives approaches that of a slow hard disk, which makes it possible to use them as an alternative form of primary storage, capable of servicing a number of audio channels. One of the major hurdles which had to be overcome in the design of such optical drives was that of making the access time suitably fast, since an optical pickup head was much more massive than the head positioner in a magnetic drive (it weighed around 100 g as opposed to less than 10 g). Techniques are being developed to rectify this situation, since it is the primary limiting factor in the onward advance of optical storage.

5.1.8 CAV and CLV modes in optical storage

CAV (constant angular velocity) and CLV (constant linear velocity) recording are two modes of rotation used in optical disk drives. In CLV recording the rotational speed of the disk changes depending on the position of the pickup, in order to keep a constant length of track passing under the head per second. In CAV recording the rotational speed of the disk remains constant. CAV disks normally have sectors of a fixed angle of arc, holding a fixed amount of data, so the data is more densely packed in sectors towards the centre of the disk (see Figure 5.8). CLV recording allows more data sectors to be stored towards the edges of the disk than at the centre, so may allow more efficient use to be made of the space available, but CLV requires servo operation to change the disk speed when the pickup head is moved, making them slower to access data.

Some drives use a mode known as zoned-CAV (Z-CAV) to pack more data into the outer tracks of a disk. The disk rotates at one of a number of fixed speeds depending on which 'zone' the pickup is in. This is really a halfway house between CAV and CLV recording and does not compromise access time so much. Compact discs use CLV recording, for example, but most optical disk cartridge drives (e.g. M-O) use a form of CAV or Z-CAV recording. Recent drives may use CAV replay, even for CLV disks, in order to enable constant spin speeds and faster access time. Z-CLV is a variant of CLV recording used on DVD-RAM disks.

5.1.9 The magneto-optical (M-O) drive

M-O drives use optical disks that can be erased and re-recorded. In order to write data, the laser is used at a higher power to that used in the reading process, to heat spots in the recording layer that is made up of rare earth elements (typically gadolinium and terbium). In older drives a biasing magnet is used to create a weak magnetic field in the vicinity of the heated spot on the disk, whose recording layer only takes on this prevailing magnetic polarisation when it is hot. Under normal conditions the recording layer cannot be magnetised (see Figure 5.9). When the spot cools it retains this magnetisation. So-called LIMDOW (light intensity modulated direct overwrite) drives have enabled better recording performance from M-O technology by doing away with the external biasing magnet. Instead the disk contains two magnetic layers with opposite polarity, close to the recording layer. The magnetic polarity taken on by the recording layer then depends on the laser intensity during recording.

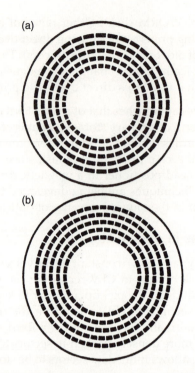

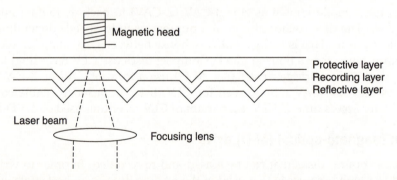

Figure 5.8 (a) Sectors on a CAV disk are of equal angle of arc. (b) On a Z-CAV disk the sector angle is not constant and more sectors are recorded at the outer edges of the disk than at the centre

Figure 5.9 The magneto-optical disk is recorded by exposing small areas of the recording layer to high-power laser light, whereupon they take on the magnetic polarity provided by the polarising magnet. On replay the magnetic polarisation affects the polarisation of reflected laser light

Although the data is recorded by a combination of optical heating and magnetisation, it is read by an entirely optical means which relies upon the fact that laser light reflected from the disk will be polarised depending on the magnetic state of the recording layer. This is known as the Kerr effect and the change in optical polarisation angle may be as small as a few degrees depending on the material concerned. The reflected light passes through a polarisation analyser, resulting in changes in intensity of the light falling on a photodetector. The M-O disk is normally pre-grooved and sectored to enable the drive to track the medium during recording.

An ISO standard was established for M-O disks, to which most of the major manufacturers adhere. This allows for two different sector sizes (512 bytes and 1024 bytes), giving 297 and 325 Mbytes per side of storage capacity respectively on a 5.25 inch disk (594 or 650 Mbytes in total) using CAV recording. There are also higher density versions offering up to around 9 Gbytes capacity, in approximate multiples of two times the basic capacity stated above.

5.1.10 Phase-change optical recording

In phase-change recording data is written by a high-powered laser, changing recorded spots from a non-crystalline (amorphous) state to a crystalline state. In the crystalline state the reflectivity is increased considerably over that of the amorphous state. Data are read by a lower-powered laser that detects changes in reflectivity. By careful selection of the recording material and laser beam control the process may be made reversible (so data may be over-written). The only apparent drawback is the number of re-write cycles allowed (cycles of erasure and re-recording), which may be in the order of ten times lower than that of the M-O disk. The CD-RW is based on phase-change principles.

5.1.11 Compact discs and drives

The CD is not immediately suitable for real-time audio editing and production, partly because of its relatively slow access time compared with hard disks, but can be seen to have considerable value for the storage and transfer of sound material that does not require real-time editing. Broadcasters use them for sound effects libraries and studios and mastering facilities use them for providing customers and record companies with 'acetates' or test pressings of a new recording. They have also become quite popular as a means of transfer-ring finished masters to a CD pressing plant in the form of the PMCD (pre-master CD). They are ideal as a means of 'proofing' CD-ROMs and other CD formats, and can be used as low-cost backup storage for computer data.

Compact discs (CDs) are familiar to most people as a consumer read-only optical disk for audio (CD-DA) or data (CD-ROM) storage. Standard audio CDs (CD-DA) conform to the Red Book standard published by Philips. The CD-ROM standard (Yellow Book) divides the CD into a structure with 2048 byte sectors, adds an extra layer of error protection, and makes it useful for general purpose data storage including the distribution of sound and video in the form of computer data files. It is possible to find disks with mixed modes, containing sections in CD-ROM format and sections in CD-Audio format. The CD Plus is one such example.

CD-R is the recordable CD, and may be used for recording CD-Audio format or other CD formats using a suitable drive and software. The Orange Book, Part 2, contains informa-tion on the additional features of CD-R, such as the area in the centre of the disk where data specific to CD-R recordings is stored. Audio CDs recorded to the Orange Book standard can be 'fixed' to give them a standard Red Book table of contents (TOC), allowing them to be replayed on any conventional CD player. Once fixed into this form, the CD-R may not subsequently be added to or changed, but prior to this there is a certain amount of flexibility, as discussed below. CD-RW disks are erasable and work on phase-change principles, requiring a drive compatible with this technology, being described in the Orange Book, Part 3.

The degree of reflectivity of CD-RW disks is much lower than that of typical CD-R and CD-ROM. This means that some early drives and players may have difficulties reading them. However the 'multi-read' specification developed by the OSTA (Optical Storage Technology Association) describes a drive that should read all types of CD, so recent drives should have no difficulties here.

Figure 5.10 shows the cross-section through a typical blank CD-R disk. The disk consists of a pre-formed 'groove' in the so-called recording layer. The recording layer consists of a green semi-transparent material, behind which is a gold reflective layer. During recording, the laser heats the recording layer to around 250 °C, a process which causes it to melt, forming a pit similar to that found on a conventional CD. On replay, the laser pickup, operated at a lower power than for recording, experiences a lower level of reflected light in the presence of a pit than it does in the absence of a pit, in exactly the same manner as for a prerecorded CD.

An Orange Book CD does not have to be recorded all at once. It can be removed from the machine and added to at a later date, appending the new material to the end of the last recording. In order to make this possible the disc contains an additional recording area inside the starting point of a conventional CD (normal CDs begin with a TOC in the centre of the disk and play from the inside out), divided into two parts (see Figure 5.11). The Program

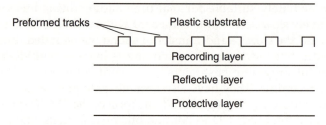

Figure 5.10 Cross section through a CD-R WORM disk

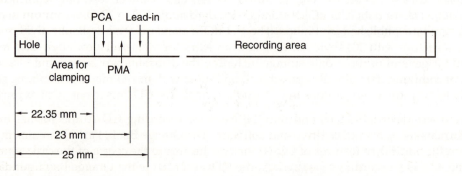

Figure 5.11 Division of recording area on the CD-R, showing space for program calibration area (PCA) and temporary program memory area (PMA)

Calibration Area (PCA) is used for optimising laser power by making a number of short test recordings when a new disk is first inserted. On subsequent occasions this calibration is not required since a message is stored on the disk to indicate the appropriate laser power. The Program Memory Area (PMA) is used to store a temporary TOC while the disk is yet 'unfixed' and this TOC is updated every time a new track is recorded. Here is also stored 'skip' information, which allows certain tracks to be skipped on replay if they have been messed up (although this will only work when the disk is replayed on a CD player that recognises skip IDs).

The lead-in area of an Orange Book CD, where a normal CD would start to read its TOC, is left blank until such time as the user decides that the disc is completed. On 'fixing' the disk the machine records a Red Book TOC, after which no further recording is allowed. The early blanks for these machines ran to 63 minutes, but 74 minute disks became available, running at the slightly slower linear velocity of $1.2 \, \mathrm{m \, s^{-1}}$. The standard capacity for a CD-R is 650 Mbyte (74 minutes), although 700 Mbyte (80 minute) disks are now available. 'Audio-only' disks have a royalty attached to them that offsets the supposed losses of the record industry owing to consumer piracy. Some consumer CD recorders may refuse to record audio on disks other than these.

A number of recording modes are possible on most Orange Book drives. 'Disk-at-once' is the most basic, in which all of the information is written at one time together with a Red Book TOC; 'Track-at-once' allows partial recording of the disk, with the option to record more at a later time, but without the option to read any of the data back until the disk TOC is fixed; 'Multisession' allows partial recording of the disk to a total of 99 sessions, with the option to read back the recorded data before the disk has been filled (provided that the reading drive is multisession capable and can read the temporary Orange Book TOC). 'Packet writing' or 'incremental writing' allows very small chunks of data to be recorded, even within a track. Only OSTA (Optical Storage Technology Association)-endorsed 'Multi-Read' CD drives can replay packet-written disks.

5.1.12 DVD

DVD is the natural successor to CD, being a higher-density optical disc format aimed at the consumer market, having the same diameter as CD and many similar physical features. It uses a different laser wavelength to CD (635–650 nm as opposed to 780 nm) so multi-standard drives need to be able to accommodate both. Data storage capacity depends on the number of sides and layers to the disk, but ranges from 4.7 Gbytes (single-layer, single-sided) up to about 18 Gbytes (double-layer, double-sided). The data transfer rate at 'one times' speed is just over $11 \, \mathrm{Mbit \, s^{-1}}$.

DVD-Video is the format originally defined for consumer distribution of movies with surround sound, typically incorporating MPEG-2 video encoding and Dolby Digital surround sound encoding. It also allows for up to eight channels of 96 kHz linear PCM audio, at up to 24-bit resolution. DVD-Audio is intended for very high quality multichannel audio reproduction and allows for linear PCM sampling rates up to 192 kHz, with numerous configurations of audio channels for different surround modes, and optional lossless data reduction (MLP). These formats will not be described in detail here as the intention is primarily to consider

Table 5.2 Recordable DVD formats

Recordable DVD type	Description
DVD-R (A and G)	DVD equivalent of CD-R. One-time recordable in sequential manner, replayable on virtually any DVD-ROM drive. Supports 'incremental writing' or 'disk at once' recording. Capacity either 3.95 (early disks) or 4.7 Gbytes per side. 'Authoring' (A) version (recording laser wavelength = 635 nm) can be used for pre-mastering DVDs for pressing, including DDP data for disk mastering (see Chapter 6). 'General' (G) version (recording laser wavelength = 650 nm) intended for consumer use, having various 'content protection' features that prevent encrypted commercial releases from being cloned
DVD-RAM	Sectored format, rather more like a hard disk in data structure when compared with DVD-R. Uses phase-change (PD-type) principles allowing direct over-write. Version 2 disks allow 4.7 Gbyte per side (reduced to about 4.2 Gbytes after formatting). Type 1 cartridges are sealed and Type 2 allow the disc to be removed. Double-sided discs only come in sealed cartridges. Can be re-written about 100 000 times. The recent Type 3 is a bare disc that can be placed in an open cartridge for recording
DVD-RW	Pioneer development, similar to CD-RW in structure, involving sequential writing. Does not involve a cartridge. Can be re-written about 1000 times. 4.7 Gbytes per side
DVD+RW	Non-DVD-Forum alternative to DVD-RAM (and not compatible), allowing direct overwrite. No cartridge. Data can be written in either CLV (for video recording) or CAV (for random access storage) modes. There is also a write-once version known as DVD+R

DVD as a mass storage medium for workstations, rather than as a consumer release format.

DVD can be used as a general-purpose data storage medium. Like CD, there are numerous different variants on the recordable DVD, partly owing to competition between the numerous different 'factions' in the DVD consortium. These include DVD-R, DVD-RAM, DVD-RW and DVD+RW, all of which are based on similar principles but have slightly different features, leading to a compatibility minefield that is only gradually being addressed. It is not proposed to go into this topic in great detail here, but a brief overview is given and a summary of common formats is shown in Table 5.2.

The 'DVD Multi' guidelines produced by the DVD Forum are an attempt to foster greater compatibility between DVD drives and disks, but this does not really solve the problem of the formats that are currently outside the DVD Forum.

Writeable DVDs are a useful option for backup of large projects, particularly DVD-RAM because of its many-times overwriting capacity and its hard disk-like behaviour. It is possible that a format like DVD-RAM could be used as primary storage in a multitrack recording/ editing system, as it has sufficient performance for a limited number of channels and it has the great advantage of being removable. However it is likely that hard disks will retain the performance edge for the foreseeable future.

5.1.13 Optical disc filing structures

There is a standard filing structure for CD-ROM known as ISO 9660 or High Sierra, which was (and still is) used when wanting to ensure that disks can be read across a wide range of platforms, although CD-ROMs can also be formatted in non-ISO modes for use on proprietary platforms. ISO 9660 format handles basic eight-character filenames and three-character extensions, but there are extensions such as 'Joliet' to allow for longer filenames.

The universal disc format (UDF) was developed as a means of simplifying the compatibility problems between optical discs such as CD and DVD, especially when used in 'packet-writing' modes. It is an IEC standard: IEC 13346. ISO 9660 compatibility is included in UDF. A form of UDF (version 1.02) was originally devised for DVD formats, version 1.5 being introduced later to encompass CD formats. It maintains 'virtual allocation tables' (VATs) on the disc that map physical data locations to relevant file packets, and these are updated to include all previous VAT data each time new packets are written.

5.1.14 Tape storage media

There are a number of types of storage media in common use for tape backup storage with AV workstations. All are cassette or cartridge formats. These include DDS, Exabyte, Mammoth, AIT and DLT. Tapes are not usually formatted in the same way as disks. Tapes are often used as basic 'data streamers' where data is stored in a very simple sequential fashion, possibly even with the block size varying in different parts of the tape. It may be that no directory is stored on the tape itself, this being kept in a disk file on the host computer. An ANSI standard exists which defines basic rules for information interchange on magnetic tapes and this is often used on media such as Exabyte to determine the method of labelling tapes and filing information. Because tapes are not usually 'mountable volumes' in the same way as disks, it is rare to be able to 'see' them on the desktops of GUI-based computers, requiring special software with appropriate drivers for the tape system in question to read and write information.

DDS is the DAT Data Storage format, and rather like the CD-ROM is the extension of a format originally intended purely for audio to general purpose data storage applications. The DAT format uses 4 mm tape and the tape is read and written using heads mounted in a drum which scans the tape in a helical fashion. On top of the audio DAT formatting is added formatting and error correction information so that the tape is then useful as a block-structured medium with low enough error rates for data purposes, and a directory area at the start of the tape.

DDS drives normally have four heads on the drum so that the data can be verified immediately after it is written – important for checking data reliability. It is recommended that one uses special DDS tapes for data purposes, which are said to be manufactured to the high specifications needed to ensure reliability, but some users have been known to use audio DAT tapes with varying degrees of success. It is sometimes necessary to alter a switch inside the drive for this purpose, so that it accepts ordinary tapes. DDS-1 drives store up to 2 Gbytes of data on a tape and some drives incorporate built-in data compression which can boost the storage capacity of such drives up to a maximum of 8 Gbytes. This is lossless compression allowing the data to be recovered in precisely its original form. The transfer rate to and from a DDS-1 drive is moderate (of the order of 180 kbyte s^{-1}), and the access time is quite slow compared with a disk drive (of the order of seconds). DDS-2 drives offer higher storage capacity and higher transfer rates. Using a longer tape, the DDS-2 drive can store up to 4 Gbytes of data in uncompressed form and up to 16 Gbytes compressed. The transfer rate is approximately 500 kbyte s^{-1}.

Exabyte tapes are based on the original consumer Video-8 format, adapted for data storage. The tapes are 8 mm wide, as opposed to the 4 mm of DDS, and the cartridges are slightly larger. Drives are typically more expensive than DDS drives. Storage capacities and transfer rates available from Exabyte drives are considerably greater than those available from DAT. One current example holds up to 5 Gbytes per tape and transfers data at a rate of around 500 kbyte s^{-1}. Maximum available capacity is currently 7 Gbytes uncompressed. Mammoth is a relatively recent tape storage technology based also on 8 mm tape. It allows considerably greater capacity than Exabyte (around 60 Gbytes) and increased data transfer rates with a simpler mechanism that is said to reduce tape wear.

The QIC (quarter-inch cartridge) is quite a well-established tape backup medium, used widely in professional computing and mainframe systems. It uses quarter-inch tape housed in a largish cartridge, and has very low error rates and high longevity. Recording is via stationary heads with multiple narrow tracks. Capacities and transfer rates are quite high, with drives storing over 10 Gbytes planned.

Digital Linear Tape (DLT) drives use a large number of linear tracks (128) across the width of a half-inch tape. It is often used for DVD masters, offering an uncompressed capacity of up to 35 Gbytes. Using a SCSI-2 interface, these drives offer transfer rates of up to 20 Mbyte s^{-1} with very low error rate, which makes them ideal for workstation backup purposes. Super DLT is a more recent alternative to DLT, offering yet higher capacity and transfer rate.

An alternative to these for high-capacity storage is AIT (Advanced Intelligent Tape), that also offers capacities into the hundreds of gigabytes and high transfer rates, as well as data compression. An interesting feature of AIT is the incorporation of a memory chip into the cassette, to store data such as a search map that enables information to be located without rewinding the tape to the directory at the start. The LTO Ultrium series of drives and cartridges, developed by HP, IBM and Seagate, has similarly high capacity and uses a 4KB cartridge memory chip that can communicate with the drive using a radio frequency transmission while the tape is not even inserted in the drive. This cartridge memory contains a file log and other user information.

5.2 Peripheral interfaces

A variety of different physical interfaces can be used for interconnecting storage devices and host workstations. Some are internal buses only designed to operate over limited lengths of cable and some are external interfaces that can be connected over several metres. The interfaces can be broadly divided into serial and parallel types, the serial types tending to be used for external connections owing to their size and ease of use. The disk interface can be slower than the drive attached to it in some cases, making it into a bottleneck in some applications. There is no point having a super fast disk drive if the interface cannot handle data at that rate.

5.2.1 SCSI

For many years the most commonly used interface for connecting mass storage media to host computers was SCSI (the Small Computer Systems Interface), pronounced 'scuzzy'. It is still used quite widely for very high performance applications but EIDE interfaces and drives are now capable of very good performance that can be adequate for many purposes.

SCSI is a high-speed parallel interface found on many computer systems, originally allowing up to seven peripheral devices to be connected to a host on a single bus. Such peripheral devices include all forms of mass storage media, CD drives, scanners, printers and network ports. It is specified in ANSI X3.131 (1986). SCSI-2 can be both faster and wider than SCSI-1, allowing for higher speed data transfer (SCSI-1 interfaces were limited to speeds of around 4–5 Mbyte s^{-1}, and were only 8 bits wide, whereas SCSI-2 can run at over 10 Mbyte s^{-1} and may be 16 or even 32 bits wide). SCSI has grown through a number of improvements and revisions, the latest being Ultra160 SCSI, capable of addressing 16 devices at a maximum data rate of 160 Mbyte s^{-1}.

SCSI devices are connected in a 'daisy-chain' fashion, as shown in Figure 5.12. SCSI-1 devices have two 50-pin connectors for this purpose, although some computers like the Macintosh have a non-standard 25-pin D-type connector. SCSI-2 usually uses a higher density connector. SCSI devices all have a means of setting their address, either with a DIP switch, a rotary or push button switch, and this determines the address on which the device will respond. The highest numbered address has the highest priority on the bus and will be dealt with first, which helps when two devices conflict in attempting to access the bus. Normally the host computer has the highest address (ID7), leaving ID0 through ID6 for peripherals. A computer's internal hard disk often uses ID0. It is important to ensure that all devices on the bus have different addresses, otherwise problems arise, although it is not necessary to assign SCSI IDs in sequence.

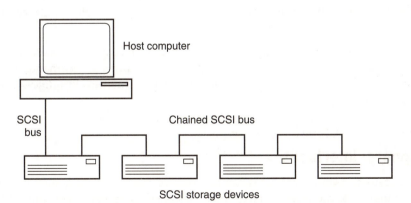

Figure 5.12 Interconnection of SCSI devices

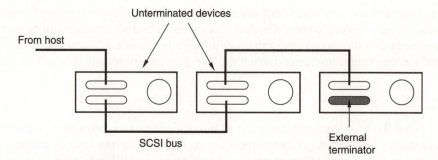

Figure 5.13 Termination of a SCSI chain, showing use of an external terminator on the last device in the chain

The SCSI bus requires termination at both ends (one end is normally in the host computer or card and is not modifiable). This termination is a collection of resistors connected to each of the parallel lines that ensure the termination impedance of the bus is correct, in order that the data is not distorted by reflections or attenuated. Unterminated SCSI buses occasionally work, but it is not recommended. Termination can be either internal or external to the peripheral and it may be switchable or automatically sensed and controlled. Internal unswitchable termination is not advisable because it forces one to use the terminated device at the end of the SCSI chain (see Figure 5.13). It is particularly inconvenient if more than one SCSI device is to be connected, because the termination has to be physically removed from those devices in the middle of the chain (not always easy). External termination normally involves plugging a termination block into the daisy-chain connector of the last device in the chain. These can be easily purchased from computer stores. Automatic termination is useful because it means that the user does not need to think about which devices are in which positions on the bus – the device senses the impedance of the bus and terminates or not accordingly. Only the devices at each end of the bus should be terminated, not any of those in between.

'The shorter the better' is the motto when it comes to choosing cables. Data rates are very high on the SCSI bus and it is important to limit cable lengths to less than a metre where possible, otherwise errors will arise. Poor quality cables are the root of many problems encountered with SCSI buses and trouble-free operation depends on using high quality cables that are double-screened.

The most common problems to arise involve (a) computers failing to 'see' certain peripherals; (b) systems failing to boot up properly; (c) data errors resulting in erroneous file transfers; (d) system crashes and 'glitches'. The following hints form a first-level troubleshooting guide:

- Never connect or disconnect SCSI devices with power turned on.
- Check that all devices have different addresses.
- Check all cables and connectors for soundness.
- Try swapping cables around or changing cables.
- Try shorter cables.
- Check termination and change if necessary.
- Try putting devices in different physical positions in the chain.

- Try changing the order of SCSI addresses.
- Try powering up SCSI devices in a different order.
- Try moving devices apart physically.
- Ensure that the correct device drivers are installed on the host computer.
- Run a SCSI diagnostic software tool which may point to the fault.

5.2.2 ATA/IDE interface

The ATA and IDE family of interfaces has evolved through the years as the primary internal interface for connecting disk drives to PC system buses. It is cheap and ubiquitous. Although drives with such interfaces were not considered adequate for audio purposes in the past, many people are now using them with the on-board audio processing of modern computers as they are cheap and the performance is adequate for many needs. It is a development of IBM's Advanced Technology Attachment (ATA) interface and has gone through improvements such as EIDE (Enhanced IDE) which improved performance and allowed four devices to be attached instead of the two of basic IDE. IDE started off using direct cylinder and sector addressing of disk drives but recent versions use logical addressing because of the disk size limitations imposed by physical addressing. Recent flavours of this interface family include Ultra ATA/66 and Ultra ATA/100 that use a 40-pin, 80-conductor connector and deliver data rates up to either 66 or 100 Mbyte s^{-1}. ATAPI (ATA Packet Interface) is a variant used for storage media such as CD drives.

Serial ATA is a relatively recent development designed to enable disk drives to be interfaced serially, thereby reducing the physical complexity of the interface. High data transfer rates are planned, eventually up to 600 Mbyte s^{-1}. It is intended primarily for internal connection of disks within host workstations, rather than as an external interface like USB or Firewire.

5.2.3 PCMCIA

PCMCIA is a standard expansion port for notebook computers and other small-size computer products. A number of storage media and other peripherals are available in PCMCIA format, and these include flash memory cards, modem interfaces and super-small hard disk drives. The standard is of greatest use in portable and mobile applications where limited space is available for peripheral storage.

5.2.4 IEEE 1394 (Firewire) and USB

Firewire and USB are both serial interfaces for connecting external peripherals. They are covered in Chapter 4 and Chapter 6 in relation to their use for MIDI and audio information respectively, so they will not be described in detail here. It is sufficient to explain that they both enable disk drives to be connected in a very simple manner, with high transfer rates (many hundreds of megabits per second), although USB 1.0 devices are limited to 12 Mbit s^{-1}. A key feature of these interfaces is that they can be 'hot plugged' (in other words devices can be connected and disconnected with the power on). The interfaces also supply basic power that enables some simple devices to be powered from the host device. Interconnection cables can usually be run up to between 5 and 10 metres, depending on the cable and the data rate.

5.3 Filing systems and volume partitions

So far only the physical structure and basic format of mass storage have been described. The way in which this raw storage space is used is another issue. There are a number of ways of organising the storage capacity of a disk drive which involve formatting it at a high level for a particular filing system, depending on the computer platform or other host device and its operating system. It is this that determines whether the files stored on a disk or tape will be accessible by the host computer, once interfaced correctly. If physical media are to be exchanged between systems, for example, then the filing system must be able to be handled by the host computer's operating system. Driver or extension software can often be obtained to enable computers to read filing systems other than their own.

When a disk is formatted at a low level the sector headers are written and the bad blocks mapped out. A map is kept of the locations of bad blocks so that they may be avoided in sub-sequent storage operations. Low level formatting can take quite a long time as every block has to be addressed. During a high level format the disk may be subdivided into a number of 'partitions'. Each of these partitions can behave as an entirely independent 'volume' of information, as if it were a separate disk drive (see Figure 5.14). It may even be possible to format each partition in a different way, such that a different filing system may be used for each partition. Each volume then has a directory created, which is an area of storage set aside to contain information about the contents of the disk. The directory indicates the locations of the files, their sizes, and various other vital statistics.

A number of audio workstation manufacturers developed their own filing systems that were optimised for speed and efficiency in real-time applications. In many cases this was the key

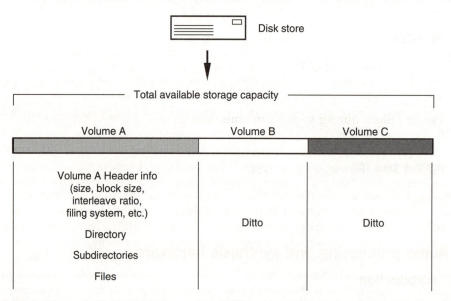

Figure 5.14 A disk may be divided up into a number of different partitions, each acting as an independent volume of information

to their success because it allowed them to obtain more simultaneous audio channels from a given disk than would otherwise have been possible. Now that disk drives have become cheap and fast, the need for special filing systems has become less important while compatibility has become more important because of the need to interchange data between systems. The most common general purpose filing systems in audio workstations are HFS (Hierarchical Filing System) or HFS+ (for Mac OS), FAT 32 (for Windows PCs) and NTFS (for Windows NT and 2000). The Unix operating system is used on some multi-user systems and high-powered workstations and also has its own filing system. These were not designed principally with real-time requirements such as audio and video replay in mind but they have the advantage that disks formatted for a widely used filing system will be more easily interchangeable than those using proprietary systems.

5.4 Formatting, fragmentation and optimisation of media

The process of formatting a disk or tape erases all of the information in the volume. (It may not actually do this, but it rewrites the directory and volume map information to make it seem as if the disk is empty again.) Effectively the volume then becomes virgin territory again and data can be written anywhere.

When an erasable volume like a hard disk has been used for some time there will be a lot of files on the disk, and probably a lot of small spaces where old files have been erased. New files must be stored in the available space and this may involve splitting them up over the remaining smaller areas. This is known as disk fragmentation, and it seriously affects the overall performance of the drive. The reason is clear to see from Figure 5.15. More head seeks are required to access the blocks of a file than if they had been stored contiguously, and this slows down the average transfer rate considerably. It may come to a point where the drive is unable to supply data fast enough for the purpose.

There are only two solutions to this problem: one is to reformat the disk completely (which may be difficult, if one is in the middle of a project), the other is to optimise or consolidate the storage space. Various software utilities exist for this purpose, whose job is to consolidate all the little areas of free space into fewer larger areas. They do this by juggling the blocks of files between disk areas and temporary RAM – a process that often takes a number of hours. Power failure during such an optimisation process can result in total corruption of the drive, because the job is not completed and files may be only half moved, so it is advisable to back up the drive before doing this. It has been known for some such utilities to make the files unusable by some audio editing packages, because the software may have relied on certain files being in certain physical places, so it is wise to check first with the manufacturer.

5.5 Audio processing and synthesis hardware

5.5.1 Introduction

A lot of audio processing now takes place within the workstation, usually relying either on the host computer's processing power (using the CPU to perform signal processing operations) or

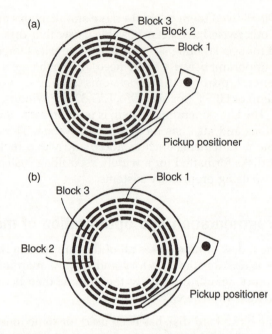

Figure 5.15 At (a) a file is stored in three contiguous blocks and these can be read sequentially without moving the head. At (b) the file is fragmented and is distributed over three remote blocks, involving movement of the head to read it. The latter read operation will take more time

on one or more DSP (digital signal processing) cards attached to the workstation's expansion bus. Professional systems usually use external A/D and D/A convertors, connected to a 'core' card attached to the computer's expansion bus. This is because it is often difficult to obtain the highest technical performance from convertors mounted on internal sounds cards, owing to the relatively 'noisy' electrical environment inside most computers. Furthermore, the number of channels required may not fit onto an internal card. As more and more audio work takes place entirely in the digital domain, though, the need for analog convertors decreases. Digital interfaces (see Chapter 6) are also often provided on external 'breakout boxes', partly for convenience and partly because of physical size of the connectors. Compact connectors such as the optical connector used for the ADAT 8-channel interface or the 2-channel SPDIF phono connector are accommodated on some cards, but multiple AES/EBU connectors cannot be.

It is also becoming increasingly common for substantial audio processing power to exist on integrated sound cards that contain digital interfaces and possibly A/D and D/A convertors. These cards are typically used for consumer or semi-professional applications on desktop computers, although many now have very impressive features and can be used for advanced operations. Such cards are now available in 'full duplex' configurations that enable audio to be received by the card from the outside world, processed and/or stored, then routed back to an external device. Full duplex operation usually allows recording and replay simultaneously.

Sound cards and DSP cards are commonly connected to the workstation using the PCI (peripheral component interface) expansion bus. Older ISA (PC) buses or NuBus (Mac) slots did not have the same data throughput capabilities and performance was therefore somewhat limited. PCI can be extended to an external expansion chassis that enables a larger number of cards to be connected than allowed for within the host computer.

Sufficient processing power can now be installed for the workstation to become the audio processing 'heart' of a larger studio system, as opposed to using an external mixing console and effects units. The higher the sampling frequency, the more DSP operations will be required per second, so it is worth bearing in mind that going up to, say, 96 kHz sampling frequency for a project will require double the processing power and twice the storage space of 48 kHz. The same is true of increasing the number of channels to which processing is applied.

5.5.2 Audio processing latency

Latency is the delay incurred in executing audio operations between input and output of a system. The lower the better is the rule, particularly when operating systems in 'full duplex' mode, because processed sound may be routed back to musicians (for foldback purposes) or may be combined with undelayed sound at some point. The management of latency is a software issue and some systems have sophisticated approaches to ensuring that all supposedly synchronous audio reaches the output at the same time no matter what processing it has encountered on the way.

Minimum latency achievable is both a hardware and a software issue. The poorest systems can give rise to tens or even hundreds of milliseconds between input and output whereas the best reduce this to a few milliseconds. Audio I/O that connects directly to an audio processing card can help to reduce latency, otherwise the communication required between host and various cards can add to the delay. Some real-time audio processing software also implements special routines to minimise and manage critical delays and this is often what distinguishes professional systems from cheaper ones. The audio driver software or 'middleware' that communicates between applications and sound cards influences latency considerably. One example of such middleware intended for low latency audio signal routing in computers is Steinberg's ASIO (Audio Stream Input Output), discussed further in Section 5.9.

5.5.3 DSP cards

DSP cards can be added to widely used workstation packages such as Digidesign's ProTools. These so-called 'DSP Farms' or 'Mix Farms' are expansion cards that connect to the PCI bus of the workstation and take on much of the 'number crunching' work involved in effects processing and mixing. 'Plug-in' processing software is becoming an extremely popular and cost-effective way of implementing effects processing within the workstation, and this is discussed further in Chapter 7. ProTools plug-ins usually rely either on DSP Farms or on host-based processing (see Section 5.5.4) to handle this load.

Digidesign's TDM (Time Division Multiplex) architecture is a useful example of the way in which audio processing can be handled within the workstation. Here the processing tasks are shared between DSP cards, each card being able to handle a certain number of operations per second. If the user runs out of 'horse power' it is possible to add further DSP cards to share the load. Audio is routed and mixed at 24-bit resolution, and a common audio bus links the cards that are connected on a separate multiway ribbon cable.

5.5.4 Host-based audio processing

An alternative to using dedicated DSP cards is to use the now substantial processing capacity of a typical desktop workstation. The success of such 'host-based processing' obviously depends on the number of tasks that the workstation is required to undertake and this capacity may vary with time and context. It is however quite possible to use the host's own CPU to run DSP 'plug-ins' for implementing equalisation, mixing and limited effects, provided it is fast enough.

The software architecture required to run plug-in operations on the host CPU is naturally slightly different to that used on dedicated DSP cards, so it is usually necessary to specify whether the plug-in is to run on the host or on a dedicated resource such as Digidesign's TDM cards. A number of applications are now appearing, however, that enable the integration of host-based (or 'native') plug-ins and dedicated DSP such as TDM-bus cards. Audio processing that runs on the host may be subject to greater latency (input to output delay) than when using dedicated signal processing, and it obviously takes up processing power that could be used for running the user interface or other software. It is nonetheless a cost-effective option for many users that do not have high expectations of a system and it may be possible to expand the system to include dedicated DSP in the future.

5.5.5 Integrated sound cards

Integrated sound cards typically contain all the components necessary to handle audio for basic purposes within a desktop computer and may be able to operate in full duplex mode (in and out at the same time). They typically incorporate convertors, DSP, a digital interface, FM and/or wavetable synthesis engines. Optionally, they may also include some sort of I/O daughter board that can be connected to a break-out audio interface, increasing the number of possible connectors and the options for external analog conversion. Such cards also tend to sport MIDI/joystick interfaces. A typical example of this type of card is the 'SoundBlaster' series from Creative Labs.

★ Any analog audio connections are normally unbalanced and the convertors may be of only limited quality compared with the best external devices. For professional purposes it is advisable to use high quality external convertors and balanced analog audio connections.

5.5.6 Synthesis engines on sound cards

The two main approaches to synthetic sound generation on PC sound cards are FM and wavetable synthesis. In FM synthesis, as pioneered by John Chowning and developed by Yamaha, the frequency of one oscillator (or 'operator') is modulated by another oscillator or chain of oscillators, as shown in Figure 5.16. The result of this frequency modulation (FM) is the creation of a complex set of sidebands or spectral components around the fundamental or 'carrier' frequency of the last oscillator in the chain, as exemplified in Figure 5.17. Quite rich timbres can be created with only a few oscillators/operators, although advanced

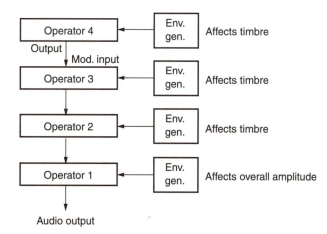

Figure 5.16 In FM synthesis one operator (the equivalent of an oscillator) frequency modulates another so as to alter its output spectrum. Each operator has its own envelope generator which affects how the output level of the operator changes with time

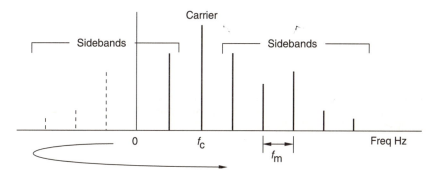

Figure 5.17 The result of FM is the creation of a sideband pattern around the 'carrier' oscillator frequency (f_c). The amplitudes and frequencies of these sidebands depend on the amplitude and frequency of the modulating signal (f_m). Sidebands are spaced apart by the frequency of the modulating signal, and a higher modulator amplitude generally creates more sidebands (a richer timbre). Sidebands which fall into the negative frequency range (below 0 Hz) are folded back into the positive range, some with phase reversal

FM synthesisers often use up to six operators per voice. The operators can be arranged in different ways, either in a chain with each modulating the next, or partly in parallel with each sub-chain contributing a particular component of the voice. These configurations are called 'algorithms', and some examples are shown in Figure 5.18. Each operator can be affected by an envelope generator which controls the way in which the amplitude of the output changes with time, and a simple envelope has four stages, as shown in Figure 5.19.

Although FM is a very flexible way of producing new synthesised sounds it is not always easy to predict or program, so as to produce a particular desired output. Wavetable synthesis is more predictable in this respect, since it involves the storage of short portions of sampled

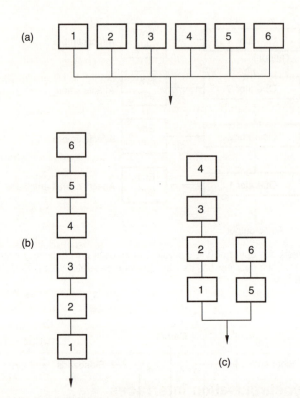

Figure 5.18 Some examples of FM synthesis algorithms. (a) Operators in parallel give the equivalent of additive synthesis (not really FM). (b) Operators in series produce very complex and unpredictable timbres. (c) A combination of series and parallel operators can be used so that different components of the sound can be handled by different parts of the algorithm

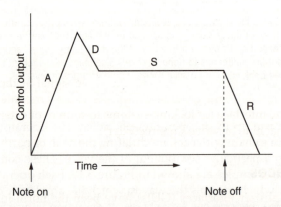

Figure 5.19 A typical envelope generator has four stages. Attack (A), Decay (D), Sustain (S) and Release (R). The rates and maximum values of each stage of the envelope can be set independently

sound waves in memory (the wave table is basically the series of memory addresses containing the discrete sample values of the stored wave segment). During replay, the wave samples are read out of memory in various ways, very similar to the replay of ordinary digital audio recordings, except that the pitch of the stored sound is varied by skipping samples in order to change the period of the replayed sound. Using variable rate replay and digital filtering a simple stored wave segment can be transformed in both pitch and timbre. A technique known as looping is used to allow quite short stored wave segments to be lengthened or sustained on replay by repeating one section of the stored wave over and over. There is a clear trade off here between the shortness of the looped segment (which conserves memory) and the quality of the instrumental sound produced. As with other forms of synthesis, envelope generators are used to alter the characteristics of the output over the duration of each note, and often separate wavetables are used for attack and sustain portions of a note.

Wavetables stored in ROM are inflexible because only a standard sound set is available to the multimedia sound designer. Increasingly common is for sound cards to contain an area of working RAM, in addition to standard sounds on ROM, which can be used for the storage of temporary or custom wavetables, now possible through the use of downloadable sound (DLS) as described in Chapter 4. In this way the sound designer can ensure that the sounds he wants to use are available when his product is played, by arranging for the appropriate audio samples to be uploaded into the sound card RAM before the action begins. New sounds could be loaded during the course of a game or other multimedia production.

5.6 External synchronisation interfaces

It is quite rare to find a means of externally synchronising the audio sampling rate of a basic sound card or computer audio system. Silicon Graphics included this as one of the digital audio options on its Unix workstations, but the average PC sound card can only be internally clocked. More sophisticated sound cards or additional hardware is usually needed to allow workstations to be externally synchronised. For example Digidesign's 'Sync I/O' interface enables ProTools to be connected to a range of sync sources and destinations. Creamware also has a 'Sync Plate' card that accompanies its SCOPE Fusion hardware and software, carrying word clock and an ADAT sync interface.

The minimum requirement for external synchronisation is usually a word clock input and output, being a BNC connector carrying a square wave signal at the sampling frequency. More advanced systems will also provide composite video sync interfaces and possibly a SMPTE/EBU timecode interface for locking systems to sources of external timecode (such as from an audio tape machine or video tape recorder).

5.7 User interfaces

Screen-based graphical user interfaces (GUIs) have been the norm in most low-cost audio workstation packages for many years because they are cheap and can display a lot of information. A user would typically interact with such an interface using a mouse and a QWERTY keyboard, which tends to limit the flexibility and controllability for complicated mixes. Dedicated audio workstations, however, have typically used a combination of screen display and physical controls. Dedicated systems are typically more expensive than PC- or Mac-based systems and have lost ground commercially as the power of desktop workstations has increased and the cost fallen. The versatility of plug-in architectures and standard computing platforms has proven very popular. However, as the power of desktop systems has increased the difficulty of controlling all the functions has also increased, as has the number of channels, leading to a need for dedicated control interfaces again. Furthermore, now that the audio workstation has the processing power to act as the mixing console and effects processor in a studio, some users are doing away with an external mixer altogether and attaching a sophisticated control interface to the workstation.

Examples of external control interfaces from Digidesign are shown in Figure 5.20. Devices such as these can range from low-cost devices to substantial control surfaces with transport control, editing control, metering and multiple moving faders.

5.8 Serial control interfaces

A variety of external serial interfaces can be provided to enable interconnection with control devices and data networks. Networking is discussed in Chapter 6. RS-232, RS-422 and MIDI are discussed in this section.

5.8.1 RS-232 and RS-422

RS-232 and RS-422 should be mentioned here as they are serial data interfaces still used quite a lot for controlling external equipment. RS-232 normally terminates on a 25-way D-type connector and carries receive and transmit data lines as well as all sorts of control and housekeeping lines for managing the flow of data between devices such as computers and modems. Data transmission is unbalanced. RS-422 normally terminates in a 9-pin D-type connector and can transfer data at a number of rates using a standard asynchronous serial communication protocol. The electrical interface is normally balanced and the collection of control and housekeeping lines that accompany RS-232 are drastically reduced to 'request to send (RTS)' and 'clear to send (CTS)' handshaking. A widely used protocol on this interface is 'Sony 9-pin' which is used a lot for controlling video equipment at a data rate of 38.4 kbit s^{-1}. The ES Bus was also developed as a universal remote control standard over RS-422, but it did not catch on as expected and does not seem to be particularly widely used.

5.8.2 The basic MIDI interface

The MIDI standard specifies a unidirectional serial interface running at 31.25 kbit s$^{-1}\pm1\%$. The rate was defined at a time when the clock speeds of microprocessors were typically much slower than they are today, this rate being a convenient division of the typical 1 or 2 MHz master clock rate. The rate had to be slow enough to be carried without excessive losses over simple cables and interface hardware, but fast enough to allow musical information to be transferred from one instrument to another without noticeable delays.

The hardware interface is shown in Figure 5.21. Most equipment using MIDI has three interface connectors: *IN*, *OUT*, and *THRU*. The OUT connector carries data that the device itself has generated. The IN connector receives data from other devices and the THRU connector is a direct throughput of the data that is present at the IN. As can be seen from the hardware

(a)

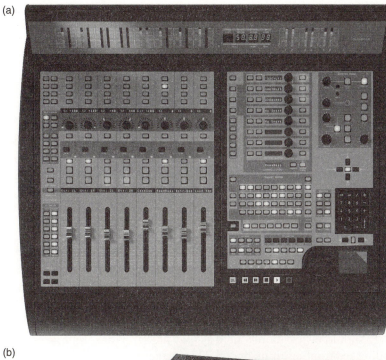

(b)

Figure 5.20 Examples of two external control interfaces from Digidesign. (a) ProControl. (b) Control 24

interface diagram, it is simply a buffered feed of the input data, and it has not been processed in any way. A few cheaper devices do not have THRU connectors, but it is possible to obtain 'MIDI THRU boxes' which provide a number of 'THRUs' from one input. Occasionally, devices without a THRU socket allow the OUT socket to be switched between OUT and THRU functions. A 5 mA current loop is created between a MIDI OUT or THRU and a MIDI IN, when

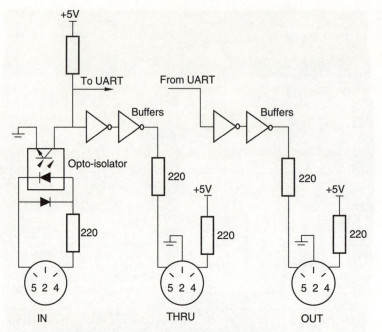

Figure 5.21 MIDI electrical interface showing IN, OUT and THRU ports

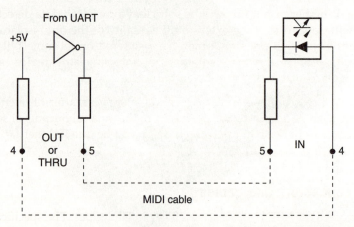

Figure 5.22 A current loop is formed between the OUT of the transmitter and the IN of the receiver when a MIDI cable is connected. The LED in the receiver's opto-isolator is turned on and off according to current flow

connected with the appropriate cable, and data bits are signalled by the turning on and off of this current by the sending device. This principle is shown in Figure 5.22.

The interface incorporates an opto-isolator between the MIDI IN (that is the receiving socket) and the device's microprocessor system. This is to ensure that there is no direct electrical link between devices and helps to reduce the effects of any problems that might occur if one

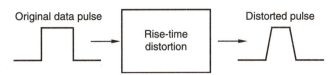

Figure 5.23 The edges of a square pulse subjected to rise-time distortion

instrument in a system were to develop an electrical fault. An opto-isolator is an encapsulated device in which a light-emitting diode (LED) can be turned on or off depending on the voltage applied across its terminals, illuminating a photo-transistor which consequently conducts or not, depending on the state of the LED. Thus the data is transferred optically, rather than electrically. In the MIDI specification, the opto-isolator is defined as having a rise time of no more than 2 μs. The rise time affects the speed with which the device reacts to a change in its input and if slow will tend to distort the leading edge of data bit cells, as shown in Figure 5.23. The same also applies in practice to fall times.

Rise-time distortion results in timing instability of the data, since it alters the time at which a data edge crosses the decision point between one and zero. If the rise time is excessively slow the data value may be corrupted since the output of the device will not have risen to its full value before the next data bit arrives. If a large number of MIDI devices are wired in series (that is from THRU to IN a number of times) the data will be forced to pass through a number of opto-isolators and thus will suffer the combined effects of a number of stages of rise-time distortion. Whether or not this will be sufficient to result in data detection errors at the final receiver will depend to some extent on the quality of the opto-isolators concerned, as well as on other losses that the signal may have suffered on its travels. It follows that the better the specification of the opto-isolator, the more stages of device cascading will be possible before unacceptable distortion is introduced.

The delay in data passed between IN and THRU is only a matter of microseconds, so this contributes little to any audible delays perceived in the musical outputs of some instruments in a large system. The bulk of any perceived delay will be due to other factors like processing delay, buffer delays and traffic.

5.8.3 MIDI connectors and cables

The connectors used for MIDI interfaces are five-pin DIN types. The specification also allows for the use of XLR-type connectors (such as those used for balanced audio signals in professional equipment), but these are rarely encountered in practice. Only three of the pins of a five-pin DIN plug are actually used in most equipment (the three innermost pins). In the cable, pin 5 at one end should be connected to pin 5 at the other, and likewise pin 4 to pin 4, and pin 2 to pin 2. Unless any hi-fi DIN cables to be used follow this convention they will not work. Professional microphone cable terminated in DIN connectors may be used as a higher-quality solution, because domestic cables will not always be a shielded twisted-pair and thus are more susceptible to external interference, as well as radiating more themselves which could interfere with adjacent audio signals.

The cable should be a shielded twisted pair with the shield connected to pin 2 of the connector at both ends, although within the receiver itself, as can be seen from Figure 5.21, the MIDI IN does not have pin 2 connected to earth. This is to avoid earth loops and makes it possible to use a cable either way round. (If two devices are connected together whose earths are at slightly different potentials, a current is caused to flow down any earth wire connecting them. This can induce interference into the data wires, possibly corrupting the data, and can also result in interference such as hum on audio circuits. It is recommended that no more than 15 m of cable is used for a single cable run in a simple MIDI system and investigation of typical cables indicates that corruption of data does indeed ensue after longer distances, although this is gradual and depends on the electromagnetic interference conditions, the quality of cable and the equipment in use. Longer distances may be accommodated with the use of buffer or 'booster' boxes that compensate for some of the cable losses and retransmit the data. It is also possible to extend a MIDI system by using a data network with an appropriate interface.

5.8.4 Interfacing a computer to a MIDI system

In order to use a workstation as a central controller for a MIDI system it must have at least one MIDI interface, consisting of at least an IN and an OUT port. (THRU is not strictly necessary in most cases.) Unless the computer has a built-in interface, as found on the old Atari machines, some form of third-party hardware interface must be added and there are many ranging from simple single ports to complex multiple port products.

A typical single port MIDI interface can be connected either to one of the spare I/O ports of the computer (a serial or USB port, for example), or can be installed as an expansion slot card (perhaps as part of an integrated sound card). Depending on which port it is connected to, some processing may be required within the MIDI interface to convert the MIDI data stream to and from the relevant interface protocol. PCs have serial interfaces that will operate at a high enough data rate for MIDI, but are not normally able to operate at precisely the 31.25 kbaud required. Nonetheless, there are a few external interfaces available which connect to the PC's serial port and transpose a higher serial data rate (often 38.4 kbaud) down to the MIDI rate using intermediate buffering and flow control. Some PCs and sound cards also have the so-called 'MIDI/joystick port' that conforms to the old Roland MPU-401 interface standard. Adaptor cables are available that provide MIDI IN and OUT connectors from this port. Some older PC interfaces also attach to the parallel port. The majority of recent MIDI interfaces are connected either to USB or Firewire ports of host workstations.

Multiport interfaces have become widely used in MIDI systems where more than 16 MIDI channels are required, and they are also useful as a means of limiting the amount of data sent or received through any one MIDI port. (A single port can become 'overloaded' with MIDI data if serving a large number of devices, resulting in data delays.) Multiport interfaces are normally more than just a parallel distribution of a single MIDI data stream, typically handling a number of independent MIDI data streams that can be separately addressed by the operating system drivers or sequencer software.

Recent interfaces are typically connected to the host workstations using USB or Firewire. On older Mac systems interconnection was handled over one or two RS-422 ports while an expansion card, RS-232 connection or parallel I/O, was normally used on the PC. The principle of such approaches is that data is transferred between the computer and the multiport interface

(a)

(b)

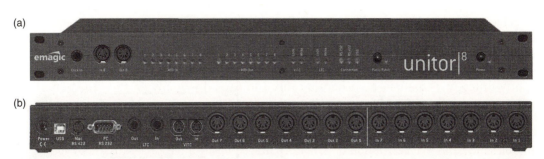

Figure 5.24 Front and back panels of the Emagic Unitor 8 interface, showing USB port, RS-422 port, RS-232 port, LTC and VITC ports and multiple MIDI ports

at a higher speed than the normal MIDI rate, requiring the interface's CPU to distribute the MIDI data between the output ports as appropriate, and transmit it at the normal MIDI rate. As described in Chapter 4, USB and Firewire MIDI protocols allow a particular stream or 'cable' to be identified so that each stream controlling 16 MIDI channels can be routed to a particular physical port or instrument.

Emagic's Unitor 8 interface is pictured in Figure 5.24. It has RS-232 and RS-422 serial ports as well as a USB port to link with the host workstation. There are eight MIDI ports with two on the front panel for easy connection of 'guest' devices or controllers that are not installed at the back. This device also has VITC and LTC timecode ports in order that synchronisation information can be relayed to and from the computer. A multi-device MIDI system is pictured in Figure 5.25, showing a number of multi-timbral sound generators connected to separate MIDI ports and a timecode connection to an external video tape recorder for use in synchronised post-production. As more of these functions are now being provided within the workstation (e.g. synthesis, video, mixing) the number of devices connected in this way will reduce.

5.9 Drivers and audio I/O software

Most audio and MIDI hardware requires 'driver' software of some sort to enable the operating system (OS) to 'see' the hardware and use it correctly. There are also sound manager or multimedia extensions that form part of the operating system of the workstation in question, designed to route audio to and from hardware in the absence of dedicated solutions. This topic crosses the boundary into software and is discussed further in Chapter 7, but basic audio and MIDI I/O extensions will be described here. (It is different from the topic of plug-in architecture which is also discussed in Chapter 7.)

The standard multimedia extensions of the OS that basic audio software used in older systems to communicate with sound cards could result in high latency and might also be limited to only two channels and 48 kHz sampling frequency. Dedicated low-latency approaches were therefore developed as an alternative, allowing higher sampling frequencies, full audio resolution, sample-accurate synchronisation and multiple channels. Examples of these are Steinberg's ASIO (Audio Stream Input Output) and Emagic's EASI. These are software extensions behaving as 'hardware abstraction layers' (HALs) that replace the

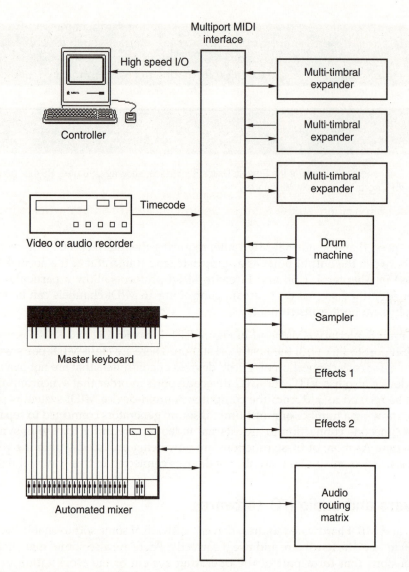

Multiport MIDI interface

High speed I/O

Controller

Timecode

Video or audio recorder

Master keyboard

Automated mixer

Multi-timbral expander

Multi-timbral expander

Multi-timbral expander

Drum machine

Sampler

Effects 1

Effects 2

Audio routing matrix

Figure 5.25 A typical multi-machine MIDI system interfaced to a computer via a multiport interface connected by a high-speed link (e.g. USB)

OS standard sound manager and enable applications to communicate more effectively with I/O hardware. ASIO, for example, handles a range of sampling frequencies and bit depths, as well as multiple channel I/O, and many sound cards and applications are ASIO-compatible.

As high-quality audio begins to feature more prominently in general purpose desktop computers, audio architectures and OS audio provision improve to keep step. OS native audio provision may now take the place of what third-party extensions have provided in the past. For example, Apple's OS X Core Audio standard is designed to provide a low-latency

HAL between applications and audio hardware, enabling multichannel audio data to be communicated to and from sound cards and external interfaces such as USB and Firewire. Core Audio handles audio in 32-bit floating-point form for high-resolution signal processing, as well as enabling sample accurate timing information to be communicated alongside audio data. Microsoft has also done something similar for Windows systems, with the Windows Driver Model (WDM) audio drivers that also include options for multichannel audio, high resolutions and sampling frequencies. DirectSound is the Microsoft equivalent of Apple's OS X Core Audio.

Core MIDI and DirectMusic do a similar thing for MIDI data in recent systems. Whereas previously it would have been necessary to install a third-party MIDI HAL such as OMS (Opcode's Open Music System) or MIDI Manager to route MIDI data to and from multiport interfaces and applications, these features are now included within the operating system's multimedia extensions.

Useful websites

The Optical Storage Technology Association: www.osta.org
DVD Forum: www.dvdforum.com
DVD+RW Alliance: www.dvdrw.com
WDM Audio: www.microsoft.com/hwdev/tech/audio/wdmaudio.asp
OS X Core Audio: http://developer.apple.com/audio/coreaudio.html

6 Audio formats and data interchange

This chapter is concerned with common formats for the storage and interchange of digital audio. It includes coverage of the most widely encountered audio and edit list file formats, digital interfaces and networking protocols.

6.1 Audio file formats

6.1.1 Introduction

There used to be almost as many file formats for audio as there are days in the year. In the computer games field, for example, this is still true to some extent. For a long time the specific file storage strategy used for disk-based digital audio was the key to success in digital workstation design, because disk drives were relatively slow and needed clever strategies to ensure that they were capable of handling a sufficiently large number of audio channels. Manufacturers also worked very much in isolation and the size of the market was relatively small, leading to virtually every workstation or piece of software using a different file format for audio and edit list information.

There are still advantages in the use of filing structures specially designed for real-time applications such as audio and video editing, because one may obtain better performance from a disk drive in this way, but the need is not as great as it used to be. Interchange is becoming at least as important as, if not more important than ultimate transfer speed and the majority of hard disk drives available today are capable of replaying many channels of audio in real time without a particularly fancy storage strategy. Indeed a number of desktop systems simply use the native filing structure of the host computer (see Chapter 5). As the use of networked workstations grows, the need for files to be transferred between systems also grows and either by international standardisation or by sheer force of market dominance certain file formats are becoming the accepted means by which data are exchanged. This is not to say that we will only be left with one or two formats, but that systems will have to be able to read and write files in the common formats if users are to be able to share work with others.

The recent growth in the importance of metadata (data about data), and the representation of audio, video and metadata as 'objects', has led to the development of interchange methods that are based on object-oriented concepts and project 'packages' as opposed to using simple text files and separate media files. There is increasing integration between audio and other media in multimedia authoring and some of the file formats mentioned below are closely related to international efforts in multimedia file exchange.

It is not proposed to attempt to describe all of the file formats in existence, because that would be a relatively pointless exercise and would not make for interesting reading. It is nonetheless useful to have a look at some examples taken from the most commonly encountered file formats, particularly those used for high quality audio by desktop and multimedia systems, since these are amongst the most widely used in the world and are often handled by audio workstations even if not their native format. It is not proposed to investigate the large number of specialised file formats developed principally for computer music on various platforms, nor the files used for internal sounds and games on many computers.

6.1.2 File formats in general

A data file is simply a series of data bytes formed into blocks and stored either contiguously or in fragmented form. In a sense files themselves are independent of the operating system and filing structure of the host computer, because a file can be transferred to another platform and still exist as an identical series of data blocks. It is the filing system that is often the platform- or operating-system-dependent entity. There are sometimes features of data files that relate directly to the operating system and filing system that created them, these being fairly fundamental features, but they do not normally prevent such files being translated by other platforms.

For example, there are two approaches to byte ordering: the so-called little-endian order in which the least significant byte comes first or at the lowest memory address, and the big-endian format in which the most significant byte comes first or at the highest memory address. These relate to the byte ordering used in data processing by the two most common microprocessor families and thereby to the two most common operating systems used in desktop audio workstations. Motorola processors, as used in the Apple Macintosh, deal in big-endian byte ordering, and Intel processors, as used in MS-DOS machines, deal in little-endian byte ordering. It is relatively easy to interpret files either way around but it is necessary to know that there is a need to do so if one is writing software.

Secondly, Macintosh files may have two parts – a resource fork and a data fork, as shown in Figure 6.1, whereas Windows files only have one part. High level 'resources' are stored in the resource fork (used in some audio files for storing information about the file, such as signal processing to be applied, display information and so forth) whilst the raw data content of the file is stored in the data fork (used in audio applications for audio sample data). The resource fork is not always there, but may be. The resource fork can get lost when transferring such files between machines or to servers, unless Mac-specific protocols are used (e.g. MacBinary or BinHex).

Some data files include a 'header', that is a number of bytes at the start of the file containing information about the data that follows (see Figure 6.2). In audio systems this may include the sampling rate and resolution of the file. Audio replay would normally be started immediately after the header. On the other hand, some files are simply raw data, usually in cases where the format is fixed. ASCII text files are a well known example of raw data files – they simply begin with the first character of the text. More recently file structures have been developed that are really 'containers' for lots of smaller files, or data objects, each with its own descriptors and data. The RIFF structure, described in Section 6.1.6, is an early example of the concept of a 'chunk-based' file structure. Apple's Bento container structure, used in OMFI, and the container structure of AAF are more advanced examples of such an approach.

The audio data in most common high-quality audio formats is stored in twos complement form (see Chapter 2) and the majority of files are used for 16- or 24-bit data, thus employing either two or three bytes per audio sample. Eight-bit files use one byte per sample.

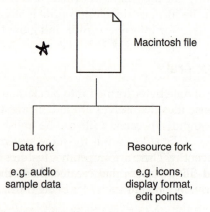

Figure 6.1 Macintosh files may have a resource and a data fork

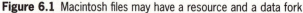

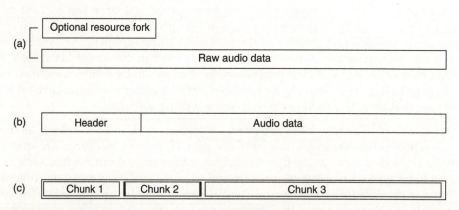

Figure 6.2 Three different kinds of sound file. (a) A simple file containing only raw audio data (showing optional Mac resource fork). (b) A file that begins with a number of bytes of header, describing the audio data that follows. (c) A chunk-format file containing self-describing chunks, each fulfilling a different function

6.1.3 Sound Designer I format

Sound Designer files originate with the Californian company Digidesign, manufacturer of probably the world's most widely used digital audio hardware for desktop computers. Many systems handle Sound Designer files because they were used widely for such purposes as the distribution of sound effects on CD-ROM and for other short music sample files. Detailed information about Digidesign file formats can be obtained if one wishes to become a third-party developer and the company exercises no particular secrecy in the matter.

The Sound Designer I format (SD I) is for mono sounds and it is recommended principally for use in storing short sounds. It originated on the Macintosh, so numerical data are stored in big-endian byte order but it has no resource fork. The data fork contains a header of 1336 bytes which is followed by the audio data bytes. The header contains information about how the sample should be displayed in Sound Designer editing software, including data describing vertical and horizontal scaling. It also contains details of 'loop points' for the file (these are principally for use with audio/MIDI sampling packages where portions of the sound are repeatedly cycled through while a key is held down, in order to sustain a note). The header contains information on the sample rate, sample period, number of bits per sample, quantisation method (e.g. 'linear', expressed as an ASCII string describing the method) and size of RAM buffer to be used.

The audio data are normally either 8- or 16-bit, and always MS byte followed by LS byte of each sample.

6.1.4 Sound Designer II format

Sound Designer II has been one of the most commonly used formats for audio workstations and has greater flexibility than SD I. Again it originated as a Mac file and unlike SD I it has a separate resource fork. The data fork contains only the audio data bytes in twos complement form, either 8 or 16 bits per sample. SD II files can contain audio samples for more than one channel, in which case the samples are interleaved, as shown in Figure 6.3, on a sample by sample basis (i.e. all the bytes for one channel sample followed by all the bytes for the next, etc.). It is unusual to find more than stereo data contained in SD II files and it is recommended that multichannel recordings are made using separate files for each channel. Some multi-channel applications, when opening stereo SD II files have first to split them into two mono

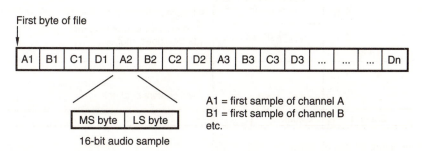

Figure 6.3 Sound Designer II files allow samples for multiple audio channels to be interleaved. Four-channel, 16-bit example shown

files before they can be used, by deinterleaving the sample data. This requires that there is sufficient free disk space for the purpose.

Since Mac resource forks can be written separately from their associated data forks, it is possible to update the descriptive information about the file separately from the audio data. This can save a lot of time (compared with single-fork files such as SD I) if the file is long and the audio data has not changed, because it saves rewriting all the audio data at the same time. Only three resources are mandatory and others can be added by developers to customise the files for their own purposes. The mandatory ones are 'sample size' (number of bytes per sample), 'sample rate' and 'channels' (describing the number of audio channels in the file). Digidesign optionally adds other resources describing things like the timecode start point and frame rates originally associated with the file, for use in post-production applications.

6.1.5 AIFF and AIFF-C formats

The AIFF format is widely used as an audio interchange standard, because it conforms to the EA IFF 85 standard for interchange format files used for various other types of information such as graphical images. AIFF is an Apple standard format for audio data and is encountered widely on Macintosh-based audio workstations and some Silicon Graphics systems. It is claimed that AIFF is suitable as an everyday audio file format as well as an interchange format and some systems do indeed use it in this way. Audio information can be stored at a number of resolutions and for any number of channels if required, and the related AIFF-C (file type 'AIFC') format allows also for compressed audio data. It consists only of a data fork, with no resource fork, making it easier to transport to other platforms.

All IFF-type files are made up of 'chunks' of data which are typically made up as shown in Figure 6.4. A chunk consists of a header and a number of data bytes to follow. The simplest AIFF files contain a 'common chunk', which is equivalent to the header data in other audio files, and a 'sound data chunk' containing the audio sample data. These are contained overall by a 'form chunk' as shown in Figure 6.5. AIFC files must also contain a 'version chunk' before the common chunk to allow for future changes to AIFC.

The common chunk header information describes the number of audio channels, the number of audio samples per channel in the following sound chunk, bits per sample (anything from

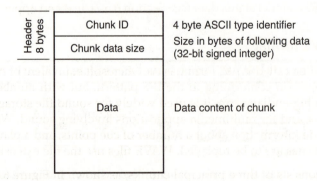

Figure 6.4 General format of an IFF file chunk

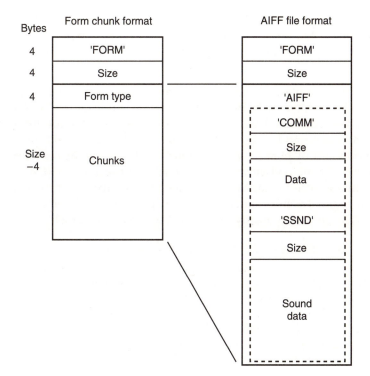

Figure 6.5 General format of an AIFF file

1 to 32), sample rate, compression type ID (AIFC only, a register is kept by Apple), and a string describing the compression type (again AIFC only). The sound data chunk consists of twos complement audio data preceded by the chunk header, the audio samples being stored as either 1, 2, 3 or 4 bytes, depending on the resolution, interleaved for multiple channels in the same way as for SD II files (see Section 6.1.4). Samples requiring less than the full eight bits of each byte should be left-justified (shifted towards the MSB), with the unused LSBs set to zero.

Optional chunks may also be included within the overall container chunk, such as marker information, comments, looping points and other information for MIDI samplers, MIDI data (see Chapter 4), AES channel status data (see Section 6.5.2), text and application-specific data.

6.1.6 RIFF WAVE format

The RIFF WAVE (often called WAV) format is the Microsoft equivalent of Apple's AIFF. It has a similar structure, again conforming to the IFF pattern, but with numbers stored in little-endian rather than big-endian form. It is used widely for sound file storage and interchange on PC workstations, and for multimedia applications involving sound. Within WAVE files it is possible to include information about a number of cue points, and a playlist to indicate the order in which the cues are to be replayed. WAVE files use the file extension '.wav'.

A basic WAV file consists of three principal chunks, as shown in Figure 6.6, the RIFF chunk, the FORMAT chunk and the DATA chunk. The RIFF chunk contains 12 bytes, the first four of which are the ASCII characters 'RIFF', the next four indicating the number of bytes in the remainder of the file (after the first eight) and the last four of which are the ASCII characters 'WAVE'. The format chunk contains information about the format of the sound file, including the number of audio channels, sampling rate and bits per sample, as shown in Table 6.1.

The audio data chunk contains a sequence of bytes of audio sample data, divided as shown in the FORMAT chunk. Unusually, if there are only 8 bits per sample or fewer each value is unsigned and ranges between 0 and 255 (decimal), whereas if the resolution is higher than this the data are signed and range both positively and negatively around zero. Audio samples are interleaved by channel in time order, so that if the file contains two channels a sample for the left channel is followed immediately by the associated sample for the right channel. The same is true of multiple channels (one sample for time-coincident sample periods on each channel is inserted at a time, starting with the lowest numbered channel), although basic WAV files were nearly always just mono or 2-channel.

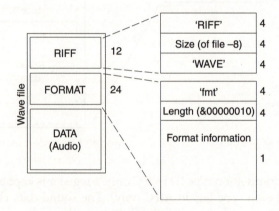

Figure 6.6 Diagrammatic representation of a simple RIFF WAVE file, showing the three principal chunks. Additional chunks may be contained within the overall structure, for example a 'bext' chunk for the Broadcast WAVE file

Table 6.1 Contents of FORMAT chunk in a basic WAVE PCM file

Byte	ID	Contents
0–3	ckID	'fmt_' (ASCII characters)
4–7	nChunkSize	Length of FORMAT chunk (binary, hex value: &00000010)
8–9	wFormatTag	Audio data format (e.g. &0001 = WAVE format PCM) Other formats are allowed, for example IEEE floating point and MPEG format (&0050 = MPEG 1)
10–11	nChannels	Number of channels (e.g. &0001 = mono, &0002 = stereo)
12–15	nSamplesPerSec	Sample rate (binary, in Hz)
16–19	nAvgBytesPerSec	Bytes per second
20–21	nBlockAlign	Bytes per sample: e.g. &0001 = 8-bit mono; &0002 = 8-bit stereo or 16-bit mono; &0004 = 16-bit stereo
22–23	nBitsPerSample	Bits per sample

The RIFF WAVE format is extensible and can have additional chunks to define enhanced functionality such as surround sound and other forms of coding. This is known as 'WAVE-format extensible'. Chunks can include data relating to cue points, labels and associated data, for example.

6.1.7 WAVE-format extensible

In order to enable the extension of the WAVE format to contain new audio formats such as certain types of surround sound and data-reduced audio (e.g. MPEG, Dolby Digital), WAVE-format extensible has a means of referring to globally unique identifiers (GUIDs) and sub-format chunks that can be vendor-specific. The 'format' chunk is extended to describe the additional content of the file, with a 'cbSize' descriptor at the end of the standard format chunk followed by the additional bytes describing the extended format. The details of this are too complex to describe here, and interested readers will find more information on the Microsoft website at www.microsoft.com/hwdev/tech/audio/multichaud.asp. The various sub-format possibilities include the option to define alternative coding formats for surround sound data that are not tied to the loudspeaker locations described below, such as B-format Ambisonic signals.

One of the necessary ambiguities to resolve was the mapping to loudspeaker locations of the multiple channels contained within a file, for speaker-feed-oriented multichannel formats. This has been achieved by defining a standard ordering of the loudspeaker locations concerned and including a channel mask word in the format chunk that indicates which channels are present. Although it is not necessary for every loudspeaker location's channel to be present in the file the samples should be presented in this order, leaving out missing channels. The first 12 of these correspond to the ordering of loudspeaker channels in the USB 1.0 specification (see Section 6.7.2), as shown in Table 6.2.

Secondly, the extensible format defines more clearly the alignment between audio sample information and the byte structure of the WAVE file, so that audio sample resolutions that do not fit exactly within a number of bytes can be handled more unambiguously. Here, wBitsPerSample must be a multiple of eight and a new field defines how many of those bits are actually used. Samples are then MSB-justified.

6.1.8 Broadcast WAVE format

The Broadcast WAVE format, described in EBU Tech. 3285, was standardised by the European Broadcasting Union (EBU) because of a need to ensure compatibility of sound files and accompanying information when transferred between workstations. It is based on the RIFF WAVE format described above, but contains an additional chunk that is specific to the format (the 'broadcast_audio_extension' chunk, ID = 'bext') and also limits some aspects of the WAVE format. Version 0 was published in 1997 and Version 1 in 2001, the only difference being the addition of a SMPTE UMID (unique material identifier) in Version 1 (this is a form of metadata). Such files currently only contain either PCM or MPEG-format audio data.

Broadcast WAVE files contain at least three chunks: the broadcast_audio_extension chunk, the format chunk and the audio data chunk. The broadcast extension chunk contains the data shown in Table 6.3. Optionally files may also contain further chunks for specialised purposes

Table 6.2 Channel ordering of WAVE format extensible audio data

Channel	Spatial location
0	Left Front (L)
1	Right Front (R)
2	Center Front (C)
3	Low Frequency Enhancement (LFE)
4	Left Surround (Ls)
5	Right Surround (Rs)
6	Left of Center (Lc)
7	Right of Center (Rc)
8	Back center (Bc)
9	Side Left (SL)
10	Side Right (SR)
11	Top (T)
12	Top Front Left (TFL)
13	Top Front Center (TFC)
14	Top Front Right (TFR)
15	Top Back Left (TBL)
16	Top Back Center (TBC)
17	Top Back Right (TBR)

Table 6.3 Broadcast_audio_extension chunk format

Data	Size (bytes)	Description
ckID	4	Chunk ID = 'bext'
ckSize	4	Size of chunk
Description	256	Description of the sound clip
Originator	32	Name of the originator
OriginatorReference	32	Unique identifier of the originator (issued by the EBU)
OriginationDate	10	'yyyy-mm-dd'
OriginationTime	8	'hh-mm-ss'
TimeReferenceLow	4	Low byte of the first sample count since midnight
TimeReferenceHigh	4	High byte of the first sample count since midnight
Version	2	BWF version number, e.g. &0001 is Version 1
UMID	64	UMID according to SMPTE 330M. If only a 32-byte UMID then the second half should be padded with zeros
Reserved	190	Reserved for extensions. Set to zero in Version 1
CodingHistory	Unrestricted	A series of ASCII strings, each terminated by CR/LF (carriage return, line feed) describing each stage of the audio coding history, according to EBU R-98

and may contain chunks relating to MPEG audio data (the 'fact' and 'mpeg_audio_extension' chunks). MPEG applications of the format are described in EBU Tech. 3285, Supplement 1 and the audio data chunk containing the MPEG data normally conforms to the MP3 frame format described in Section 6.1.9.

A multichannel extension chunk has recently been proposed for Broadcast WAVE files that defines the channel ordering, surround format, downmix coefficients for creating a two-channel mix, and some descriptive information. There are also chunks defined for metadata describing the audio contained within the file, such as the 'quality chunk' (ckID = 'qlty'), which together with the coding history contained in the 'bext' chunk make up the so-called 'capturing report'. These are described in Supplement 2 to EBU Tech. 3285. Finally there is a chunk describing the peak audio level within a file, which can aid automatic programme level setting and programme interchange.

6.1.9 MPEG audio file formats

It is possible to store MPEG-compressed audio in AIFF-C or WAVE files, with the compression type noted in the appropriate header field. There are also older MS-DOS file extensions used to denote MPEG audio files, notably .MPA (MPEG Audio) or .ABS (Audio Bit Stream). However, owing to the ubiquity of the so-called 'MP3' format (MPEG 1, Layer 3) for audio distribution on the Internet, MPEG audio files are increasingly denoted with the extension '.MP3'. Such files are relatively simple, being really no more than MPEG audio frame data in sequence, each frame being preceded by a frame header. Although the frame header at the beginning of the file might be considered to relate to all the remaining audio information, there is the possibility that settings may change during the course of replay. For example, the bit rate can change in variable bit rate modes, or joint stereo coding might be switched on, so each frame header should ideally be correctly decoded. The following describes the basic format of .MP3 files and Figure 6.7 shows the structure of a typical MPEG frame.

Layer 1 frames correspond to 384 original PCM samples (8 ms at a sampling rate of 48 kHz), and Layer 2 and 3 frames correspond to 1152 PCM samples (24 ms @ 48 kHz). The frame consists of a 32-bit header, a 16-bit CRC check word, the audio data (consisting of subband

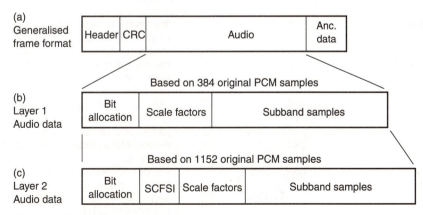

Figure 6.7 MPEG-Audio frame format

samples, appropriate scale factors and information concerning the bit allocation to different parts of the spectrum) and an ancillary data field whose length is currently unspecified. The 32-bit header of each frame consists of the information shown in Table 6.4.

MPEG files can usually be played from pretty well anywhere in the file by looking for the next frame header. Layer 1 and 2 frames are self-contained and can be decoded immediately, but Layer 3 (the only one that should strictly be called MP3) is slightly different and can take up to nine frames before the decoding can be completed correctly. This is because of the bit reservoir technique that is used to share bits optimally between a series of frames.

A so-called 'ID3 tag' has been included in many MP3 files as a means of describing the contents of the file, such as the title, artists, and so forth. This is usually found in the last 128 bits of the whole file and should not be decoded as audio. It does not begin with a frame header

Table 6.4 MPEG audio frame header

Function	No. of bits	Description
Sync word	11	All set to binary '1' to act as a synchronisation pattern. (Technically the first 12 bits were intended to be the sync word and the following ID a single bit, but see MPEG 2.5 below)
ID bits	2	Indicates the ID of the algorithm in use: '11' = MPEG 1 (ISO 11172-3); '10' = MPEG 2 (ISO 13818-3); '01' = reserved; '00' = MPEG 2.5 (an unofficial version that allowed even lower bit rates than specified in the original MPEG 2 standard, by using lower sampling frequencies (8, 11 and 12 kHz))
Layer	2	Indicates the MPEG layer in use: '11' = Layer 1; '10' = Layer 2; '01' = Layer 3; '00' = reserved
Protection bit	1	Indicates whether error correction data has been added to the audio bitstream ('0' if yes)
Bitrate index	4	Indicates the total bit rate of the channel according to a table which relates the state of these 4 bits to rates in each layer
Sampling frequency	2	Indicates the original PCM sampling frequency: '00' = 44.1 kHz; '01' = 48 kHz; '10' = 32 kHz
Padding bit	1	Indicates in state '1' that a slot has been added to the frame to make the average bit rate of the data-reduced channel relate exactly to the original sampling rate
Private bit	1	Available for private use
Mode	2	'00' = stereo; '01' = joint stereo; '10' = dual channel; '11' = single channel
Mode extension	2	Used for further definition of joint stereo coding mode to indicate either which bands are coded in joint stereo, or which type of joint coding is to be used
Copyright	1	'1' = copyright protected
Original/copy	1	'1' = original; '0' = copy
Emphasis	2	Indicates audio pre-emphasis type: '00' = none; '01' = 50/15 μs; '11' = CCITT J17

Table 6.5 A simple ID3 tag structure

Sign	Length (bytes)	Description
A	3	Tag identification. Normally ASCII 'TAG'.
B	30	Title
C	30	Artist
D	30	Album
E	4	Year
F	30	Comment string (only 28 bytes followed by '\0' in some versions)
G	1	This may represent the track number, or may be part of the comment string
H	1	Genre

Table 6.6 MPEG file ID3v2 tag header

Function	No. of bytes	Description
File identifier	3	ASCII 'ID3'
Version	2	& 03 00 (major version then revision number)
Flags	1	Binary 'abc00000' a = unsynchronisation used when set b = extended header present when set c = flag for indicating experimental version of tag
Size	4	Four lots of 0xxxxxxx, concatenated ignoring the MSB of each byte to make a 28-bit word that indicates the length of the tag after the header and including any unsynchronisation bytes

sync pattern so most software will not attempt to decode it as audio. A simple tag has the typical format shown in Table 6.5.

ID3v2.2 is a much more developed and extended ID3 tagging structure for a range of information contained within frames of data, each frame having its own header. The ID3, version 2 header format is shown in Table 6.6. Patterns of data within the ID3 information that might look like an audio sync pattern are dealt with using a method known as unsynchronisation that modifies the tag to prevent the sync pattern occurring. It does this by inserting an extra zero-valued byte after the first byte of the false sync pattern.

6.1.10 DSD-IFF file format

Sony and Philips have introduced Direct Stream Digital (DSD), as already discussed in Section 2.7, as an alternative high-resolution format for audio representation. The DSD-IFF file format is based on a similar structure to other IFF-type files, described above, except that it is modified slightly to allow for the large file sizes that may be encountered. Specifically the container FORM chunk is labelled 'FRM8' and this identifies all local chunks that follow as having 'length' indications that are 8 bytes long rather than the normal 4. In other words, rather than a 4-byte chunk ID followed by a 4-byte length indication, these files have a 4-byte ID followed by an 8-byte length indication. This allows for the definition of chunks with a length greater than 2 Gbytes, which may be needed for mastering SuperAudio CDs.

In such a file, local mandatory chunks are the format version chunk ('FVER'), the property chunk ('PROP'), containing information such as sampling frequency, number of channels and loudspeaker configuration, and at least one DSD or DST (direct stream transfer – the losslessly compressed version of DSD) sound data chunk ('DSD' or 'DST'). There are also various optional chunks that can be used for exchanging more detailed information and comments such as might be used in project interchange. Further details of this file format, and an excellent guide to the use of DSD-IFF in project applications, can be found in the DSD-IFF specification, as described in the Further reading at the end of this chapter.

6.1.11   Edit decision list (EDL) files

EDL formats have usually been unique to the workstation on which they are used but the need for open interchange is increasing the pressure to make EDLs transportable between packages. There is an old and widely used format for EDLs in the video world that is known as the CMX-compatible form. CMX is a well-known manufacturer of video editing equipment and most editing systems will read CMX EDLs for the sake of compatibility. These can be used for basic audio purposes, and indeed a number of workstations can read CMX EDL files for the purpose of auto-conforming audio edits to video edits performed on a separate system. The CMX list defines the cut points between source material and the various transition effects at joins, and it can be translated reasonably well for the purpose of defining audio cut points and their timecode locations, using SMPTE/EBU form, provided video frame accuracy is adequate.

Software can be obtained for audio and video workstations that translates EDLs between a number of different standards to make interchange easier, although it is clear that this process is not always problem-free and good planning of in-house processes is vital. The OMFI structure also contains a format for interchanging edit list data, as described below. AES 31 (see below) is now gaining considerable popularity among workstation software manufacturers as a simple means of exchanging audio editing projects between systems. The Advanced Authoring Format (AAF) is becoming increasingly relevant to the exchange of media project data between systems, and is likely to take over from OMFI as time progresses.

6.1.12   AES 31 format

AES 31 is an international standard designed to enable straightforward interchange of audio files and projects between systems. Audio editing packages are increasingly offering AES 31 as a simple interchange format for edit lists. In Part 1 the standard specifies a disk format that is compatible with the FAT32 file system, a widely used structure for the formatting of computer hard disks. Part 2 is not finalised at the time of writing but is likely to describe an audio file format closely based on the Broadcast WAVE format. Part 3 describes simple project interchange, including a format for the communication of edit lists using ASCII text that can be parsed by a computer as well as read by a human. The basis of this is the edit decision markup language (EDML). It is not necessary to use all the parts of AES 31 to make a satisfactory interchange of elements. For example, one could exchange an edit list according to part 3 without using a disk based on part 1. Adherence to all the parts would mean that one could take a removable disk from one system, containing sound files and a project file, and the project would be readable directly by the receiving device.

EDML documents are limited to a 7-bit ASCII character set in which white space delimits fields within records. Standard carriage return (CR) and line-feed (LF) characters can be included to aid the readability of lists but they are ignored by software that might parse the list. An event location is described by a combination of timecode value and sample count information. The timecode value is represented in ASCII using conventional hours, minutes, seconds and frames (e.g. HH:MM:SS:FF) and the optional sample count is a four-figure number denoting the number of samples after the start of the frame concerned at which the event actually occurs. This enables sample-accurate edit points to be specified. It is slightly more complicated than this because the ASCII delimiters between the timecode fields are changed to indicate various parameters:

`HH:MM` delimiter = `Frame count and timebase indicator` (see Table 6.7)

`MM:SS` delimiter = `Film frame indicator (if not applicable, use the previous delimiter)`

`SS:FF` delimiter = `Video field and timecode type` (see Table 6.8)

The delimiter before the sample count value is used to indicate the audio sampling frequency, including all the pull-up and pull-down options (e.g. f_s times $1/1.001$). There are too many of

Table 6.7 Frame count and timebase indicator coding in AES 31

Frame count	Unknown	Timebase	
		1.000	*1.001*
30	?	|	:
25	!	.	/
24	#	=	-

Table 6.8 Video field and timecode type indicator in AES 31

Counting mode	Video field	
	Field 1	*Field 2*
PAL	.	:
NTSC non-drop-frame	.	:
NTSC drop-frame	,	;

these possibilities to list here and the interested reader is referred to the standard for further information. This is an example of a timecode and (after the slash denoting 48 kHz sampling frequency) optional sample count value:

```
14:57:24.03/0175
```

The Audio Decision List (ADL) is contained between two ASCII keyword tags <ADL> and </ADL>. It in turn contains a number of sections, each contained within other keyword tags such as <VERSION>, <PROJECT>, <SYSTEM> and <SEQUENCE>. The edit points themselves are contained in the <EVENT_LIST> section. Each event begins with the ASCII keyword '(Entry)', which serves to delimit events in the list, followed by an entry number (32-bit integer, incrementing through the list) and an entry type keyword to describe the nature of the event (e.g. '(Cut)'). Each different event type then has a number of bytes following that define the event more specifically. The following is an example of a simple cut edit, as suggested by the standard:

```
(Entry)  0010  (Cut)  F  "FILE://VOL/DIR/FILE"  1  1  03:00:00;00/0000
01:00:00:00/0000  01:00:10:00/0000_
```

This sequence essentially describes a cut edit, entry number 0010, the source of which is the file (F) with the path shown, using channel 1 of the source file (or just a mono file), placed on track 1 of the destination timeline, starting at timecode three hours in the source file, placed to begin at one hour in the destination timeline (the 'in point') and to end ten seconds later (the 'out point'). Some workstation software packages store a timecode value along with each sound file to indicate the nominal start time of the original recording (e.g. BWF files contain a timestamp in the 'bext' chunk), otherwise each sound file is assumed to start at time zero.

It is assumed that default crossfades will be handled by the workstation software itself. Most workstations introduce a basic short crossfade at each edit point to avoid clicks, but this can be modified by 'event modifier' information in the ADL. Such modifiers can be used to adjust the shape and duration of a fade in or fade out at an edit point. There is also the option to point at a rendered crossfade file for the edit point, as described in Chapter 3.

6.1.13 The Open Media Framework Interchange (OMFI)

OMFI was introduced in 1994 by Avid Technology, an American company specialising in desktop audio and video post-production systems (now merged with Digidesign). It was an attempt to define a common standard for the interchange of audio, video, edit list and other multimedia information between workstations running on different platforms. It was in effect a publicly available format and Avid did not charge licensing fees of any kind, OMFI being a means of trying to encourage greater growth in this field of the industry as a whole. A number of other manufacturers signed up to support OMF and worked jointly on its development. Avid makes available an OMF Interchange Toolkit at moderate cost for developers who want to build OMF compatibility into their products. The company is gradually migrating from OMFI to a new format called AAF (Advanced Authoring Format) that is supported by a wide range of multimedia vendors. Parts of OMFI 2.0 have apparently been incorporated within AAF.

The OMFI 1.0 specification was lengthy and dealt with descriptions of the various types of information that could be contained and the methods of containment. It also contained details of compositions and the ways in which edit timing data should be managed. Version 1.0 of OMFI was very video oriented and specified no more for audio than the two common formats for the audio data files and a means of specifying edit points and basic crossfade durations (but not the shape). Compared with AES 31, OMFI is much more difficult for programmers to understand because it is much more expandable and versatile, being based on object-oriented concepts rather than being a simple text-based description of the project. OMFI 2.0 is yet more involved and links information in a different way to 1.0, the two being incompatible. As far as the audio user is concerned, the 1.0 version specifies that the common audio formats to be used are the uncompressed versions of either the AIFF format or the WAVE format (see above), depending on the intended hardware platform. It also allows for the possibility that manufacturers might want to specify 'private' interchange formats of their own. The format apparently limits audio resolution to 16 bits for interchange, but there is no particular reason why this should be so and some programmers have modified the Toolkit code to accommodate 24-bit audio files. Most of the version 1.0 document referred to video operations, so cuts and effects were all described in video terms.

OMFI 1.0 projects contain two types of information: 'compositions' and 'sources'. Compositions specify how the various sources are to be assembled in order to play the finished product. Source data (audio, video, or other multimedia files) may be stored either in separate files, referenced by the OMF file, or within the OMF container structure. The container structure is similar to the IFF model described above (indeed Avid originally started to use IFF), in that it contains a number of self-describing parts, and is called Bento (an Apple development). Each part of the OMFI file is complete in itself and can be handled independently of the other parts – indeed applications do not need to be able to deal with every component of an OMFI file – allowing different byte ordering for different parts if required. Systems may claim OMFI compatibility yet still not be able to deal with some of the data objects contained within the file, requiring care in implementation and some discipline in the use of OMFI within organisations. The fact that OMFI used Apple's Bento container was one of the problems encountered by the AES when attempting to standardise editing project interchange for the AES 31 standard. Since Bento is not an open standard and is Apple's proprietary technology, AES apparently could not adopt OMFI directly.

6.1.14 MXF – the Media Exchange Format

MXF was developed by the Pro-MPEG forum as a means of exchanging audio, video and metadata between devices, primarily in television operations. It is based on the modern concept of media objects that are split into 'essence' and 'metadata'. Essence files are the raw material (i.e. audio and video) and the metadata describes things about the essence (such as where to put it, where it came from and how to process it).

MXF files attempt to present the material in a 'streaming' format, that is one that can be played out in real time, but they can also be exchanged in conventional file transfer operations. As such they are normally considered to be finished program material, rather than material that is to be processed somewhere downstream, designed for playout in broadcasting environments. The bit stream is also said to be compatible with recording on digital videotape devices.

6.1.15 AAF – the Advanced Authoring Format

AAF is an authoring format for multimedia data that is supported by numerous vendors, including Avid which has adopted it as a migration path from OMFI (see above). Parts of OMFI 2.0 form the basis for parts of AAF and there are also close similarities between AAF and MXF (described in the previous section). Like the formats to which it has similarities, AAF is an object-oriented format that combines essence and metadata within a container structure. Unlike MXF it is designed for project interchange such that elements within the project can be modified, post-processed and resynchronised. It is not, therefore, directly suitable as a streaming format but can easily be converted to MXF for streaming if necessary.

Rather like OMFI it is designed to enable complex relationships to be described between content elements, to map these elements onto a timeline, to describe the processing of effects, synchronise streams of essence, retain historical metadata and refer to external essence (essence not contained within the AAF package itself). It has three essential parts: the AAF Object Specification (which defines a container for essence and metadata, the logical contents of objects and rules for relationships between them); the AAF Low-Level Container Specification (which defines a disk filing structure for the data, based on Microsoft's Structured Storage); and the AAF SDK Reference Implementation (which is a software development kit that enables applications to deal with AAF files). The Object Specification is extensible in that it allows new object classes to be defined for future development purposes.

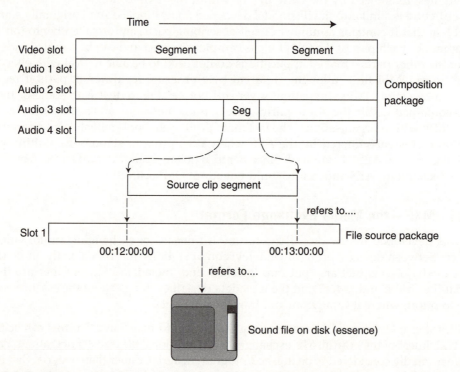

Figure 6.8 Graphical conceptualisation of some metadata package relationships in AAF: a simple audio post-production example

The basic object hierarchy is illustrated in Figure 6.8, using an example of a typical audio post-production scenario. 'Packages' of metadata are defined that describe either compositions, essence or physical media. Some package types are very 'close' to the source material (they are at a lower level in the object hierarchy, so to speak) – for example a 'file source package' might describe a particular sound file stored on disk. The metadata package, however, would not be the file itself, but it would describe its name and where to find it. Higher level packages would refer to these lower level packages in order to put together a complex program. A composition package is one that effectively describes how to assemble source clips to make up a finished program. Some composition packages describe effects that require a number of elements of essence to be combined or processed in some way.

Packages can have a number of 'slots'. These are a bit like tracks in more conventional terminology, each slot describing only one kind of essence (e.g. audio, video, graphics). Slots can be static (not time-dependent), timeline (running against a timing reference) or event-based (one-shot, triggered events). Slots have segments that can be source clips, sequences, effects or fillers. A source clip segment can refer to a particular part of a slot in a separate essence package (so it could refer to a short portion of a sound file that is described in an essence package, for example).

6.2 Disk pre-mastering formats

The original tape format for submitting CD masters to pressing plants was Sony's audio-dedicated PCM 1610/1630 format on U-matic video tape. This is now 'old technology' and has been replaced by alternatives based on more recent data storage media and file storage protocols. These include the PMCD (pre-master CD), CD-R, Exabyte and DLT tape formats. DVD mastering also requires high-capacity media for transferring the many gigabytes of information to mastering houses in order that glass masters can be created.

The Disk Description Protocol (DDP) developed by Doug Carson and Associates is now widely used for describing disk masters. Version 1 of the DDP laid down the basic data structure but said little about higher level issues involved in interchange, making it more than a little complicated for manufacturers to ensure that DDP masters from one system would be readable on another. Version 2 addressed some of these issues.

DDP is a protocol for describing the contents of a disk, which is not medium specific. That said it is common to interchange CD masters with DDP data on 8 mm Exabyte data cartridges and DVD masters are typically transferred on DLT Type III or IV compact tapes or on DVD-R(A) format disks with CMF (cutting master format) DDP headers. DDP files can be supplied separately to the audio data if necessary. DDP can be used for interchanging the data for a number of different disk formats, such as CD-ROM, CD-DA, CD-I and CD-ROM-XA, DVD-Video and Audio, and the protocol is really extremely simple. It consists of a number of 'streams' of data, each of which carries different information to describe the contents of the disk. These streams may be either a series of packets of data transferred over a network, files on a disk or tape, or raw blocks of data independent of any filing system. The DDP protocol simply maps its data into whatever block or packet size is used by the medium concerned, provided that the block or packet size is at least 128 bytes. Either a standard computer filing structure can be used, in which case each stream is contained within a named file, or the storage medium is used 'raw' with each stream starting at a designated sector or block address.

The ANSI tape labelling specification is used to label the tapes used for DDP transfers. This allows the names and locations of the various streams to be identified. The principal streams included in a DDP transfer for CD mastering are as follows:

1 DDP ID stream or 'DDPID' file. 128 bytes long, describing the type and level of DDP information, various 'vital statistics' about the other DDP files and their location on the medium (in the case of physically addressed media), and a user text field (not transferred to the CD).
2 DDP Map stream or 'DDPMS' file. This is a stream of 128-byte data packets which together give a map of the CD contents, showing what types of CD data are to be recorded in each part of the CD, how long the streams are, what types of subcode are included, and so forth. Pointers are included to the relevant text, subcode and main streams (or files) for each part of the CD.
3 Text stream. An optional stream containing text to describe the titling information for volumes, tracks or index points (not currently stored in CD formats), or for other text comments. If stored as a file, its name is indicated in the appropriate map packet.
4 Subcode stream. Optionally contains information about the subcode data to be included within a part of the disk, particularly for CD-DA. If stored as a file, its name is indicated in the appropriate map packet.
5 Main stream. Contains the main data to be stored on a part of the CD, treated simply as a stream of bytes, irrespective of the block or packet size used. More than one of these files can be used in cases of mixed-mode disks, but there is normally only one in the case of a conventional audio CD. If stored as a file, its name is indicated in the appropriate map packet.

6.3 Interconnecting audio devices

In the case of analog interconnection between devices, replayed digital audio is converted to the analog domain by the replay machine's D/A convertors, routed to the recording machine via a conventional audio cable and then reconverted to the digital domain by the recording machine's A/D convertors. The audio is subject to any gain changes that might be introduced by level differences between output and input, or by the record gain control of the recorder and the replay gain control of the player. Analog domain copying is necessary if any analog processing of the signal is to happen in between one device and another, such as gain correction, equalisation, or the addition of effects such as reverberation. Most of these operations, though, are now possible in the digital domain.

An analog domain copy cannot be said to be a perfect copy or a clone of the original master, since the data values will not be exactly the same (owing to slight differences in recording level, differences between convertors, the addition of noise, and so on). For a clone it is necessary to make a true digital copy.

Professional digital audio systems, and some consumer systems, have digital interfaces conforming to one of the standard protocols and allow for a number of channels of digital audio data to be transferred between devices with no loss of sound quality. Any number of generations of digital copies may be made without affecting the sound quality of the latest generation, provided that errors have been fully corrected. The digital outputs of a recording device are taken from a point in the signal chain after error correction, which results in the copy being error corrected. Thus the copy does not suffer from any errors that existed in the master, provided that those errors were correctable. This process takes place in real time, requiring the operator to put the receiving device into record mode such that it simply stores the incoming stream of audio data. Any accompanying metadata may or may not be recorded (often most of it is not).

Digital interfaces may be used for the interconnection of recording systems and other audio devices such as mixers and effects units. It is now common only to use analog interfaces at the very beginning and end of the signal chain, with all other interconnections being made digitally.

Making a copy of a recording using any of the digital interface standards involves the connection of appropriate cables between player and recorder, and the switching of the recorder's input to 'digital' as opposed to 'analog', since this sets it to accept a signal from the digital input as opposed to the A/D convertor. It is necessary for both machines to be operating at the same sampling frequency (unless a sampling frequency convertor is used) and may require the recorder to be switched to 'external sync' mode, so that it can lock its sampling frequency to that of the player. Alternatively (and preferably) a common reference signal may be used to synchronise all devices that are to be interconnected digitally. A recorder should be capable of at least the same quantising resolution (number of bits per sample) as a player, otherwise audio resolution will be lost. If there is a difference in resolution between the systems it is advisable to use a processor in between the machines that optimally dithers the signal for the new resolution, or alternatively to use redithering options on the source machine to prepare the signal for its new resolution.

6.4 Computer networks and digital audio interfaces compared

Dedicated 'streaming' interfaces, as employed in broadcasting, production and post-production environments, are the digital audio equivalent of analog signal cables, down which signals for one or more channels are carried in real time from one point to another, possibly with some auxiliary information (metadata) attached. An example is the AES-3 interface, described below. Such an audio interface uses a data format dedicated to audio purposes, whereas a computer data network may carry numerous types of information.

Dedicated interfaces are normally unidirectional, point-to-point connections, and should be distinguished from buses and computer networks that are often bidirectional and carry data in a packet format. Sources may be connected to destinations using a routeing matrix or by patching individual connections, very much as with analog signals. Audio data are transmitted in an unbroken stream, there is no handshaking process involved in the data transfer, and erroneous data are not retransmitted because there is no mechanism for requesting its retransmission. The data rate of a dedicated audio interface is directly related to the audio sampling frequency, wordlength and number of channels of the audio data to be transmitted, ensuring that the interface is always capable of serving the specified number of channels. If a channel is unused for some reason its capacity is not normally available for assigning to other purposes (such as higher-speed transfer of another channel, for example).

Dedicated audio interfaces, therefore, may be thought of as best suited to operational situations in which analog signal cabling needs to be replaced by a digital equivalent, and where digital audio signals are to be routed from place to place within a studio environment so as to ensure dedicated signal feeds. There are, however, a number of developments in real-time computer networking that begin to blur the distinction between such approaches and conventional asynchronous file transfers, owing to the increased use of 'streaming media', as discussed below.

There is an increasing trend towards employing standard computer interfaces and networks to transfer audio information, as opposed to using dedicated audio interfaces. Such computer networks are typically used for a variety of purposes in general data communications and they may need to be adapted for audio applications that require sample-accurate real-time transfer. The increasing ubiquity of computer systems in audio environments makes it inevitable that generic data communication technology will gradually take the place of dedicated interfaces. It also makes sense economically to take advantage of the 'mass market' features of the computer industry.

Computer networks are typically general-purpose data carriers that may have asynchronous features and may not always have the inherent quality-of-service (QoS) features that are required for 'streaming' applications. They also normally use an addressing structure that enables packets of data to be carried from one of a number of sources to one of a number of destinations and such packets will share the connection in a more or less controlled way. Data transport protocols such as TCP/IP are often used as a universal means of managing the transfer of data from place to place, adding overheads in terms of data rate, delay and error handling that may work against the efficient transfer of audio. Such networks may be intended primarily for file transfer applications where the time taken to transfer the file is not a crucial factor – as fast as possible will do.

Conventional office Ethernet is a good example of a computer network interface that has limitations in respect of audio streaming. The original 10 Mbit s^{-1} data rate was quite slow, although theoretically capable of handling a number of channels of real-time audio data. If employed between only two devices and used with a low-level protocol such as UDP (user datagram protocol) audio can be streamed quite successfully, but problems can arise when multiple devices contend for use of the bus and where the network is used for general purpose data communications in addition to audio streaming. There is no guarantee of a certain quality of service, because the bus is a sort of 'free for all', 'first-come-first-served' arrangement that is not designed for real-time applications. To take a simple example, if one's colleague attempts to download a huge file from the Internet just when one is trying to stream a broadcast live-to-air in a local radio station, using the same data network, the chances are that one's broadcast will drop out occasionally.

One can partially address such limitations in a crude way by throwing data-handling capacity at the problem, hoping that increasing the network speed to 100 Mbit s^{-1} or even 1 Gbit s^{-1} will avoid it ever becoming overloaded. Circuit-switched networks can also be employed to ease these problems (that is networks where individual circuits are specifically established between sources and destinations). Unless capacity can be reserved and service quality guaranteed a data network will never be a suitable replacement for dedicated audio interfaces in critical environments such as broadcasting stations. This has led to the development of real-time protocols and/or circuit-switched networks for handling audio information on data interfaces, in which latency (delay) and bandwidth are defined and guaranteed. The audio industry can benefit from the increased data rates, flexibility and versatility of general purpose interfaces provided that these issues are taken seriously.

Desktop computers and consumer equipment are also increasingly equipped with general purpose serial data interfaces such as USB (universal serial bus) and FireWire (IEEE 1394). These are examples of personal area network (PAN) technology, allowing a number of devices to be interconnected within a limited range around the user. These have a high enough data rate to carry a number of channels of audio data over relatively short distances, either over copper or optical fibre. Audio protocols also exist for these, as described below.

6.5 Dedicated audio interface formats

6.5.1 Digital interface types

There are a number of types of digital interface, some of which are international standards and others of which are manufacturer-specific. They all carry digital audio for one or more channels with at least 16-bit resolution and will operate at the standard sampling rates of 44.1 and 48 kHz, as well as at 32 kHz if necessary, some having a degree of latitude for varispeed. Some interface standards have been adapted to handle higher sampling frequencies such as 88.2 and 96 kHz. The interfaces vary as to how many physical interconnections are required. Some require one link per channel plus a synchronisation signal, whilst others carry all the audio information plus synchronisation information over one cable.

The interfaces are described below in outline. It is common for subtle incompatibilities to arise between devices, even when interconnected with a standard interface, owing to the different ways in which non-audio information is implemented. This can result in anything from minor operational problems to total non-communication and the causes and remedies are unfortunately far too detailed to go into here. The reader is referred to *The Digital Interface Handbook* by Rumsey and Watkinson, as well as to the standards themselves, if a greater understanding of the intricacies of digital audio interfaces is required.

6.5.2 The AES 3 interface (AES 3)

The AES 3 interface, described almost identically in AES3-1992, IEC 60958 and EBU Tech. 3250E among others, allows for two channels of digital audio (A and B) to be transferred serially over one balanced interface, using drivers and receivers similar to those used in the RS422 data transmission standard, with an output voltage of between 2 and 7 volts as shown in Figure 6.9. The interface allows two channels of audio to be transferred over distances up to 100 m, but longer distances may be covered using combinations of appropriate cabling, equalisation and termination. Standard XLR-3 connectors are used, often labelled DI (for digital in) and DO (for digital out).

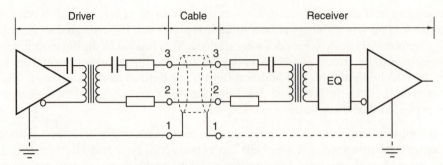

Figure 6.9 Recommended electrical circuit for use with the standard two-channel interface

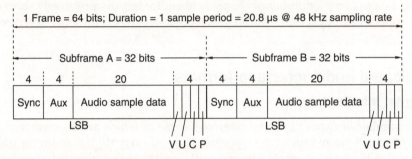

Figure 6.10 Format of the standard two-channel interface frame

Each audio sample is contained within a 'subframe' (see Figure 6.10), and each subframe begins with one of three synchronising patterns to identify the sample as either A or B channel, or to mark the start of a new channel status block (see Figure 6.11). These synchronising patterns violate the rules of bi-phase mark coding (see below) and are easily identified by a decoder. One frame (containing two audio samples) is normally transmitted in the time period of one audio sample, so the data rate varies with the sampling frequency. (Note, though, that the recently introduced 'single-channel-double-sampling-frequency' mode of the interface allows two samples for one channel to be transmitted within a single frame in order to allow the transport of audio at 88.2 or 96 kHz sampling frequency.)

Additional data is carried within the subframe in the form of 4 bits of auxiliary data (which may either be used for additional audio resolution or for other purposes such as low-quality speech), a validity bit (V), a user bit (U), a channel status bit (C) and a parity bit (P), making 32 bits per subframe and 64 bits per frame. Channel status bits are aggregated at the receiver to form a 24-byte word every 192 frames, and each bit of this word has a specific function relating to interface operation, an overview of which is shown in Figure 6.12. Examples of bit usage in this word are the signalling of sampling frequency and pre-emphasis, as well as the carrying of a sample address 'timecode' and labelling of source and destination. Bit 1 of the first byte signifies whether the interface is operating according to the professional (set to 1) or consumer (set to 0) specification.

Bi-phase mark coding, the same channel code as used for SMPTE/EBU timecode, is used in order to ensure that the data is self-clocking, of limited bandwidth, DC free, and polarity

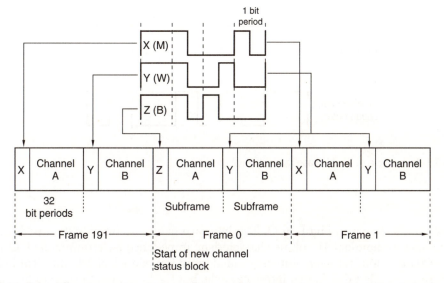

Figure 6.11 Three different preambles (X, Y and Z) are used to synchronise a receiver at the starts of subframes

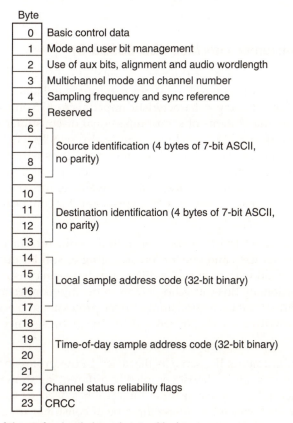

Byte	
0	Basic control data
1	Mode and user bit management
2	Use of aux bits, alignment and audio wordlength
3	Multichannel mode and channel number
4	Sampling frequency and sync reference
5	Reserved
6	Source identification (4 bytes of 7-bit ASCII, no parity)
7	
8	
9	
10	Destination identification (4 bytes of 7-bit ASCII, no parity)
11	
12	
13	
14	Local sample address code (32-bit binary)
15	
16	
17	
18	Time-of-day sample address code (32-bit binary)
19	
20	
21	
22	Channel status reliability flags
23	CRCC

Figure 6.12 Overview of the professional channel status block

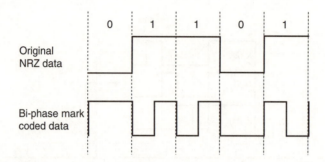

Figure 6.13 An example of the bi-phase mark channel code

independent, as shown in Figure 6.13. The interface has to accommodate a wide range of cable types and a nominal 110 ohms characteristic impedance is recommended. Originally (AES3-1985) up to four receivers with a nominal input impedance of 250 ohms could be connected across a single professional interface cable, but the later modification to the standard recommended the use of a single receiver per transmitter, having a nominal input impedance of 110 ohms.

6.5.3 Standard consumer interface (IEC 60958-3)

The most common consumer interface (historically related to SPDIF – the Sony/Philips digital interface) is very similar to the AES 3 interface, but uses unbalanced electrical interconnection over a coaxial cable having a characteristic impedance of 75 ohms, as shown in Figure 6.14. It can be found on many items of semi-professional or consumer digital audio equipment, such as CD players and DAT machines, and is also widely used on computer sound cards because of the small physical size of the connectors. It usually terminates in an RCA phono connector, although some equipment makes use of optical fibre interconnects (TOSlink) carrying the same data. Format convertors are available for converting consumer format signals to the professional format, and vice versa, and for converting between electrical and optical formats.

When the IEC standardised the two-channel digital audio interface, two requirements existed: one for 'consumer use', and one for 'broadcasting or similar purposes'. A single IEC standard (IEC 958) resulted with only subtle differences between consumer and professional implementation. Occasionally this caused problems in the interconnection of machines, such as when consumer format data was transmitted over professional electrical interfaces. IEC 958 has now been rewritten as IEC 60958 and many of these uncertainties have been addressed.

The data format of subframes is the same as that used in the professional interface, but the channel status implementation is almost completely different, as shown in Figure 6.15. The second byte of channel status in the consumer interface has been set aside for the indication of 'category codes', these being set to define the type of consumer usage. Current examples of defined categories are (00000000) for the General category, (10000000) for Compact Disc

Driver
Cable
Receiver

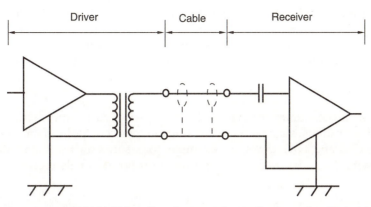

Figure 6.14 The consumer electrical interface (transformer and capacitor are optional but may improve the electrical characteristics of the interface)

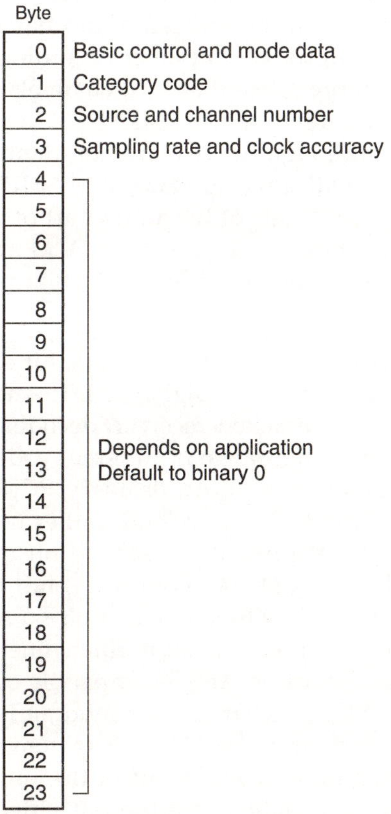

Byte
0 Basic control and mode data
1 Category code
2 Source and channel number
3 Sampling rate and clock accuracy
4
5
6
7
8
9
10
11
12
13
14
15
16
17
18
19
20
21
22
23
Depends on application
Default to binary 0

Figure 6.15 Overview of the consumer channel status block

and (11000000) for a DAT machine. Once the category has been defined, the receiver is expected to interpret certain bits of the channel status word in a particular way, depending on the category. For example, in CD usage, the four control bits from the CD's 'Q' channel subcode are inserted into the first four control bits of the channel status block (bits 1–4). Copy protection can be implemented in consumer-interfaced equipment, according to the Serial Copy Management System (SCMS).

The user bits of the consumer interface are often used to carry information derived from the subcode of recordings, such as track identification and cue point data. This can be used when copying CDs and DAT tapes, for example, to ensure that track start ID markers are copied along with the audio data. This information is not normally carried over AES/EBU interfaces.

6.5.4 Carrying data-reduced audio over standard digital interfaces

The increased use of data-reduced multichannel audio has resulted in methods by which such data can be carried over standard two-channel interfaces, either for professional or consumer purposes. This makes use of the 'non-audio' or 'other uses' mode of the interface, indicated in the second bit of channel status, which tells conventional PCM audio decoders that the information is some other form of data that should not be converted directly to analog audio. Because data-reduced audio has a much lower rate than the PCM audio from which it was derived, a number of audio channels can be carried in a data stream that occupies no more space than two channels of conventional PCM. These applications of the interface are described in SMPTE 337M (concerned with professional applications) and IEC 61937, although the two are not identical. SMPTE 338M and 339M specify data types to be used with this standard. The SMPTE standard packs the compressed audio data into 16, 20 or 24 bits of the audio part of the AES 3 sub-frame and can use the two sub-frames independently (e.g. one for PCM audio and the other for data-reduced audio), whereas the IEC standard only uses 16 bits and treats both sub-frames the same way.

Consumer use of this mode is evident on DVD players, for example, for connecting them to home cinema decoders. Here the Dolby Digital or DTS-encoded surround sound is not decoded in the player but in the attached receiver/decoder. IEC 61937 has parts, either pending or published, dealing with a range of different codecs including ATRAC, Dolby AC-3, DTS and MPEG (various flavours). An ordinary PCM convertor trying to decode such a signal would simply reproduce it as a loud, rather unpleasant noise, which is not advised and does not normally happen if the second bit of channel status is correctly observed. Professional applications of the mode vary, but are likely to be increasingly encountered in conjunction with Dolby E data reduction – a relatively recent development involving mild data reduction for professional multichannel applications in which users wish to continue making use of existing AES 3-compatible equipment (e.g. VTRs, switchers and routers). Dolby E enables 5.1-channel surround audio to be carried over conventional two-channel interfaces and through AES 3-transparent equipment at a typical rate of about 1.92 Mbit s^{-1} (depending on how many bits of the audio sub-frame are employed). It is designed so that it can be switched or edited at video frame boundaries without disturbing the audio.

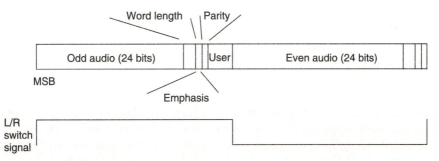

Figure 6.16 Format of TDIF data and LRsync signal

6.5.5 Tascam digital interface (TDIF)

Tascam's interfaces have become popular owing to the widespread use of the company's DA-88 multitrack recorder and more recent derivatives. The primary TDIF-1 interface uses a 25-pin D-sub connector to carry eight channels of audio information in two directions (in and out of the device), sampling frequency and pre-emphasis information (on separate wires, two for f_s and one for emphasis) and a synchronising signal. The interface is unbalanced and uses CMOS voltage levels. Each data connection carries two channels of audio data, odd channel and MSB first, as shown in Figure 6.16. As can be seen, the audio data can be up to 24 bits long, followed by 2 bits to signal the word length, 1 bit to signal emphasis and 1 bit for parity. There are also four user bits per channel that are not usually used.

This resembles a modified form of the AES3 interface frame format. An accompanying left/right clock signal is high for the odd samples and low for the even samples of the audio data. It is difficult to find information about this interface but the output channel pairs appear to be on pins 1–4 with the left/right clock on pin 5, while the inputs are on pins 13–10 with the left/right clock on pin 9. Pins 7, 14–17 (these seem to be related to output signals) and 22–25 (related to the input signals) are grounded. The unbalanced, multi-conductor, non-coaxial nature of this interface makes it only suitable for covering short distances up to 5 metres.

6.5.6 Alesis digital interface

The ADAT multichannel optical digital interface, commonly referred to as the 'light pipe' interface or simply 'ADAT Optical', is a serial, self-clocking, optical interface that carries eight channels of audio information. It is described in US Patent 5,297,181: 'Method and apparatus for providing a digital audio interface protocol'. The interface is capable of carrying up to 24 bits of digital audio data for each channel and the eight channels of data are combined into one serial frame that is transmitted at the sampling frequency. The data is encoded in NRZI format for transmission, with forced ones inserted every five bits (except during the sync pattern) to provide clock content. This can be used to synchronise the sampling clock of a receiving device if required, although some devices require the use of a separate 9-pin ADAT sync cable for synchronisation. The sampling frequency is normally limited to 48 kHz with varispeed up to 50.4 kHz and TOSLINK optical connectors are typically employed (Toshiba TOCP172 or equivalent). In order to operate at 96 kHz sampling frequency some

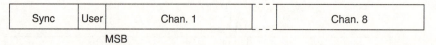

Sync	User	Chan. 1		Chan. 8

MSB

Figure 6.17 Basic format of ADAT data

implementations use a 'double-speed' mode in which two channels are used to transmit one channel's audio data (naturally halving the number of channels handled by one serial interface). Although 5 m lengths of optical fibre are the maximum recommended, longer distances may be covered if all the components of the interface are of good quality and clean. Experimentation is required.

As shown in Figure 6.17 the frame consists of an 11-bit sync pattern consisting of 10 zeros followed by a forced one. This is followed by four user bits (not normally used and set to zero), the first forced one, then the first audio channel sample (with forced ones every five bits), the second audio channel sample, and so on.

6.5.7 Roland R-bus

Roland has recently introduced its own proprietary multichannel audio interface that, like TDIF (but not directly compatible with it), uses a 25-way D-type connector to carry eight channels of audio in two directions. Called R-bus it is increasingly used on Roland's digital audio products and convertor boxes are available to mediate between R-bus and other interface formats. Little technical information about R-bus is available publicly at the time of writing.

6.5.8 Sony digital interface for DSD (SDIF-3)

Sony has recently introduced a high-resolution digital audio format known as 'Direct Stream Digital' or DSD (see Chapter 2). This encodes audio using one-bit sigma-delta conversion at a very high sampling frequency of typically 2.8224 MHz (64 times 44.1 kHz). There are no internationally agreed interfaces for this format of data, but Sony has released some preliminary details of an interface that can be used for the purpose, known as SDIF-3. Some early DSD equipment used a data format known as 'DSD-raw' which was simply a stream of DSD samples in non-return-to-zero (NRZ) form, as shown in Figure 6.18(a).

In SDIF-3 data is carried over 75 ohm unbalanced coaxial cables, terminating in BNC connectors. The bit rate is twice the DSD sampling frequency (or 5.6448 Mbit s^{-1} at the sampling frequency given above) because phase modulation is used for data transmission as shown in Figure 6.18(b). A separate word clock at 44.1 kHz is used for synchronisation purposes. It is also possible to encounter a DSD clock signal connection at the 64 times 44.1 kHz (2.8224 MHz).

6.5.9 Sony multichannel DSD interface (MAC-DSD)

Sony has also developed a multichannel interface for DSD signals, capable of carrying 24 channels over a single physical link. The transmission method is based on the same technology as used for the Ethernet 100BASE-TX (100 Mbit s^{-1}) twisted-pair physical layer (PHY),

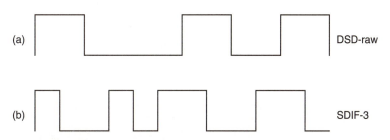

Figure 6.18 Direct Stream Digital interface data is either transmitted 'raw', as shown at (a) or phase modulated as in the SDIF-3 format shown at (b)

but it is used in this application to create a point-to-point audio interface. Category 5 cabling is used, as for Ethernet, consisting of eight conductors. Two pairs are used for bi-directional audio data and the other two pairs for clock signals, one in each direction.

Twenty-four channels of DSD audio require a total bit rate of 67.7 Mbit s^{-1}, leaving an appreciable spare capacity for additional data. In the MAC-DSD interface this is used for error correction (parity) data, frame header and auxiliary information. Data is formed into frames that can contain Ethernet MAC headers and optional network addresses for compatibility with network systems. Audio data within the frame is formed into 352 32-bit blocks, 24 bits of each being individual channel samples, six of which are parity bits and two of which are auxiliary bits.

In a recent enhancement of this interface, Sony has introduced 'SuperMAC' which is capable of handling either DSD or PCM audio with very low latency (delay), typically less than 50 μs. The number of channels carried depends on the sampling frequency. Twenty-four DSD channels can be handled, or 48 PCM channels at 44.1/48 kHz, reducing proportionately as the sampling frequency increases. In conventional PCM mode the interface is transparent to AES3 data including user and channel status information.

6.6 Networking

6.6.1 Basic principles of networking

A network carries data either on wire or optical fibre, and is normally shared between a number of devices and users. The sharing is achieved by containing the data in packets of a limited number of bytes (usually between 64 and 1518), each with an address attached. The packets may share a common physical link, normally a high speed serial bus of some kind, being multiplexed in time either using a regular slot structure synchronised to a system clock (isochronous transfer) or in an asynchronous fashion whereby the time interval between packets may be varied or transmission may not be regular, as shown in Figure 6.19. The length of packets may not be constant, depending on the requirements of different protocols sharing the same network. Packets for a particular file transfer between two devices may not be contiguous and may be transferred erratically, depending on what other traffic is sharing the same physical link.

Figure 6.20 shows some common physical layouts for local area networks (LANs). LANs are networks that operate within a limited area, such as an office building or studio centre,

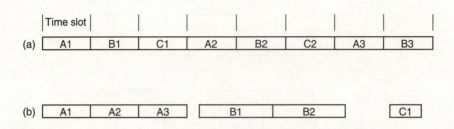

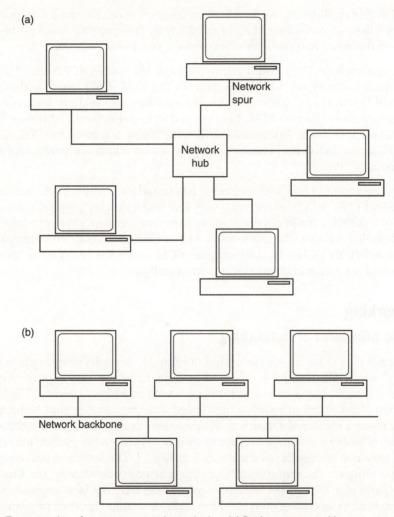

Figure 6.19 Packets for different destinations (A, B and C) multiplexed onto a common serial bus. (a) Time division multiplexed into a regular time slot structure. (b) Asynchronous transfer showing variable time gaps and packet lengths between transfers for different destinations

Figure 6.20 Two examples of computer network topologies. (a) Devices connected by spurs to a common hub, and (b) devices connected to a common 'backbone'. The former is now by far the most common, typically using CAT 5 cabling

within which it is common for every device to 'see' the same data, each picking off that which is addressed to it and ignoring the rest. Routers and bridges can be used to break up complex LANs into subnets. WANs (wide area networks) and MANs (metropolitan area networks) are larger entities that link LANs within communities or regions. PANs (personal area networks) are typically limited to a range of a few tens of metres around the user (e.g. Firewire, USB, Bluetooth). Wireless versions of these network types are increasingly common, as discussed in Section 6.6.8.

In order to place a packet of data on the network, devices must have a means for determining whether the network is busy and there are various protocols in existence for arbitrating network access. Taking Ethernet as an example: in the 'backbone' configuration devices are connected to spurs off a common serial bus that requires the bus to be 'chained' between each successive device. Here, a break in the chain can mean disconnection for more than one device. The star configuration involves a central hub or switch that distributes the data to each device separately. This is more reliable because a break in any one link does not affect the others. Bus arbitration in both these cases is normally performed by collision detection which is a relatively crude approach, relying very much on the rules of spoken conversation between people. Devices attempt to place packets on the bus whenever it appears to be quiet, but a collision may take place if another device attempts to transmit before the first one has finished. The network interface of the transmitting device detects the collision by attempting to read the data it has just transmitted and retransmits it after transmitting a brief 'blocking signal' if it has been corrupted by the collision.

A token ring configuration places each device within a 'ring', each device having both an 'in' and an 'out' port, with bus arbitration performed using a process of token passing from one device to the next. This works rather like trains running on a single track line, in that a single token is carried by the train using the line and trains can only use the line if carrying the token. The token is passed to the next train upon leaving the single-track sector to show that the line is clear. Network communication is divided into a number of 'layers', each relating to an aspect of the communication protocol and interfacing correctly with the layers either side. The ISO seven-layer model for open systems interconnection (OSI) shows the number of levels at which compatibility between systems needs to exist before seamless interchange of data can be achieved (Figure 6.21). It shows that communication begins with the application is passed down through various stages to the layer most people understand – the physical layer, or the piece of wire over which the information is carried. Layers 3, 4 and 5 can be grouped under the broad heading of 'protocol', determining the way in which data packets are formatted and transferred. There is a strong similarity here with the exchange of data on physical media, as discussed earlier, where a range of compatibility layers from the physical to the application determine whether or not one device can read another's disks.

6.6.2 Extending a network

It is common to need to extend a network to a wider area or to more machines. As the number of devices increases so does the traffic, and there comes a point when it is necessary to divide a network into zones, separated by 'repeaters', 'bridges' or 'routers'. Some of these devices allow network traffic to be contained within zones, only communicating between the zones when necessary. This is vital in large interconnected networks because otherwise data

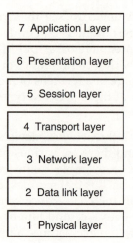

7 Application Layer

6 Presentation layer

5 Session layer

4 Transport layer

3 Network layer

2 Data link layer

1 Physical layer

Figure 6.21 The ISO model for Open Systems Interconnection is arranged in seven layers, as shown here

placed anywhere on the network would be present at every other point on the network, and overload could quickly occur.

A repeater is a device that links two separate segments of a network so that they can talk to each other, whereas a bridge isolates the two segments in normal use, only transferring data across the bridge when it has a destination address on the other side. A router is very selective in that it examines data packets and decides whether or not to pass them depending on a number of factors. A router can be programmed only to pass certain protocols and only certain source and destination addresses. It therefore acts as something of a network policeman and can be used as a first level of ensuring security of a network from unwanted external access. Routers can also operate between different standards of network, such as between FDDI and Ethernet, and ensure that packets of data are transferred over the most time/cost-effective route.

One could also use some form of router to link a local network to another that was quite some distance away, forming a wide area network (WAN), as shown in Figure 6.22. Data can be routed either over dialled data links such as ISDN (see below), in which the time is charged according to usage just like a telephone call, or over leased circuits. The choice would depend on the degree of usage and the relative costs. The Internet provides a means by which LANs are easily interconnected, although the data rate available will depend on the route, the service provider and the current traffic.

6.6.3 Network standards

Ethernet, FDDI (Fibre Distributed Data Interface), ATM (Asynchronous Transfer Mode) and Fibre Channel are examples of network standards, each of which specifies a number of layers within the OSI model. FDDI, for example, specifies only the first three layers of the OSI model (the physical, data link and network layers).

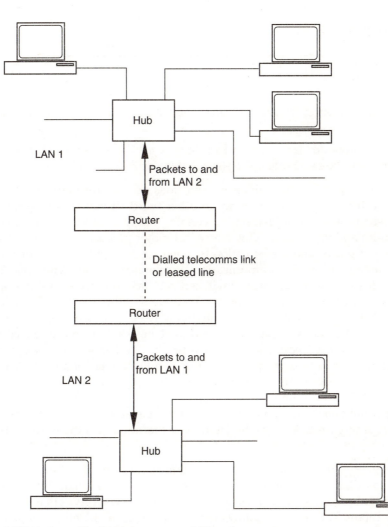

Figure 6.22 A WAN is formed by linking two or more LANs using a router

Ethernet allows a number of different methods of interconnection and runs at various rates from 10 Mbit s^{-1} to 1 Gbit s^{-1}, using collision detection for network access control. Twisted-pair (Base-T) connection using CAT 5 cabling and RJ 45 connectors is probably the most widely encountered physical interconnect these days, usually configured in the star topology using a central hub or switch. Devices can then be plugged and unplugged without affecting others on the network. Interconnection can also be via either thick (Base-10) or thin (Base-2) coaxial cable, normally working in the backbone-type configuration shown in the previous section, using 50-ohm BNC connectors and T-pieces to chain devices on the network (see Figure 6.23). Such a configuration requires resistive terminators at the ends of the bus to avoid reflections, as with SCSI connections.

FDDI is a high speed optical fibre network running at 100 Mbit s^{-1}, operating on the token passing principle described above, allowing up to 2 km between stations. It is often used as

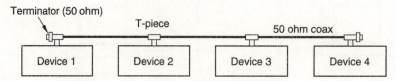

Figure 6.23 Typical thin Ethernet interconnection arrangement (this is becoming less common now)

a high-speed backbone for large networks. There is also a copper version of FDDI called CDDI which runs at the same rate but restricts interconnection distance.

ATM is a protocol for data communication and does not specify the physical medium for interconnection. It is connection-oriented, in other words it sets up connections between source and destination and can guarantee a certain quality of service, which makes it quite suitable for audio and video data. ATM allows for either guaranteed bandwidth communications between a source and a destination (needed for AV applications), or for more conventional variable bandwidth communication. It operates in a switched fashion and can extend over wide or metropolitan areas. Switched networks involve the setting up of specific circuits between a transmitter and one or more receivers, rather like a dialled telephone network (indeed this is the infrastructure of the digital telephone network). The physical network is made up of a series of interconnected switches that are reconfigured to pass the information from sources to destinations according to the header information attached to each data packet. A network management system handles the negotiation between different devices that are contending for bandwidth, according to current demand. ATM typically operates over SONET (synchronous optical network) or SDH (synchronous digital hierarchy) networks, depending on the region of the world. Data packets on ATM networks consist of a fixed 48 bits, typically preceded by a 5-byte header that identifies the virtual channel of the packet.

Fibre Channel is used increasingly for the interconnection of workstations and disk storage arrays, using so-called 'storage area network' structures. It uses a half-duplex interface, but it has a separate fibre (or copper connection) for transmit and receive circuits so can operate in full-duplex mode. The standard allows for data rates up to 4 Gbit s^{-1} depending on the capabilities of the implementation.

6.6.4 Network protocols

A protocol specifies the rules of communication on a network. In other words it determines things like the format of data packets, their header information and addressing structure, and any handshaking and error retrieval schemes, amongst other things. One physical network can handle a wide variety of protocols, and packets conforming to different protocols can coexist on the same bus.

Some common examples of general purpose network protocols are TCP/IP (Transport Control Protocol/Internet Protocol), used for communications over the Internet (see below) and also over LANs, and UDP (User Datagram Protocol) often used for basic streaming applications. These general-purpose protocols are not particularly efficient or reliable for real-time audio transfer, but they can be used for non-real-time transfer of audio files between workstations or for streaming. Specially designed protocols may be needed for audio purposes, as described below.

6.6.5 Audio network requirements

The principal application of computer networks in audio systems is in the transfer of audio data files between workstations, or between workstations and a central 'server' which stores shared files. The device requesting the transfer is known as the 'client' and the device providing the data is known as the 'server'. When a file is transferred in this way a byte-for-byte copy is reconstructed on the client machine, with the file name and any other header data intact. There are considerable advantages in being able to perform this operation at speeds in excess of real time for operations in which real-time feeds of audio are not the aim. For example, in a news editing environment a user might wish to load up a news story file from a remote disk drive in order to incorporate it into a report, this being needed as fast as the system is capable of transferring it. Alternatively, the editor might need access to remotely stored files, such as sound files on another person's system, in order to work on them separately. In audio post-production for films or video there might be a central store of sound effects, accessible by everyone on the network, or it might be desired to pass on a completed portion of a project to the next stage in the post-production process.

Wired Ethernet is fast enough to transfer audio data files faster than real time, depending on network loading and speed. For satisfactory operation it is advisable to use 100 Mbit s^{-1} or even 1 Gbit s^{-1} Ethernet as opposed to the basic 10 Mbit s^{-1} version. Switched Ethernet architectures allow the bandwidth to be more effectively utilised, by creating switched connections between specific source and destination devices. Approaches using FDDI or ATM are appropriate for handling large numbers of sound file transfers simultaneously at high speed. Unlike a real-time audio interface, the speed of transfer of a sound file over a packet-switched network (when using conventional file transfer protocols) depends on how much traffic is currently using it. If there is a lot of traffic then the file may be transferred more slowly than if the network is quiet (very much like motor traffic on roads). The file might be transferred erratically as traffic volume varies, with the file arriving at its destination in 'spurts'. There therefore arises the need for network communication protocols designed specifically for the transfer of real-time data, which serve the function of reserving a proportion of the network bandwidth for a given period of time, as described below. This is known as engineering a certain 'quality of service'.

Without real-time protocols designed as indicated above, the computer network may not be relied upon for transferring audio where an unbroken audio output is to be reconstructed at the destination from the data concerned. The faster the network the more likely it is that one would be able to transfer a file fast enough to feed an unbroken audio output, but this should not be taken for granted. Even the highest speed networks can be filled up with traffic! This may seem unnecessarily careful until one considers an application in which a disk drive elsewhere on the network is being used as the source for replay by a local workstation, as illustrated in Figure 6.24. Here it must be possible to ensure guaranteed access to the remote disk at a rate adequate for real-time transfer, otherwise gaps will be heard in the replayed audio.

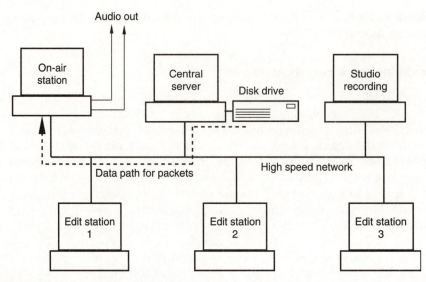

Figure 6.24 In this example of a networked system a remote disk is accessed over the network to provide data for real time audio playout from a workstation used for on-air broadcasting. Continuity of data flow to the on-air workstation is of paramount importance here

6.6.6 ISDN

ISDN is an extension of the digital telephone network to the consumer, providing two 64 kbit s^{-1} digital channels ('B' channels) which can be connected to ISDN terminals anywhere in the world simply by dialling. Data of virtually any kind may be transferred over the dialled-up link, and potential applications for ISDN include audio transfer. ISDN is really a subset of ATM, and ATM has sometimes been called broadband ISDN for this reason.

The total usable capacity of a single ISDN-2 connection is only 128 kbit s^{-1} and so it is not possible to carry linear PCM data at normal audio resolutions over such a link, but it is possible to carry moderately high-quality stereo digital audio at this rate using a data reduction system such as MPEG (see Chapter 2). Higher rates can be achieved by combining more than one ISDN link to obtain data rates of, say, 256 or 384 kbit s. Multiple ISDN lines must be synchronised together using devices known as inverse multiplexers if the different delays that may arise over different connections are to be compensated. There are also ISDN-30 lines, providing 30 simultaneous 'B' channels of 64 kbit s^{-1} (giving roughly 2 Mbit s^{-1}).

It is possible to use ISDN links for non-real time audio file transfer, and this can be economically viable depending on the importance of the project and the size of files. The cost of an ISDN call is exactly the same as the equivalent duration of normal telephone call, and therefore it can be quite a cheap way of getting information from one place to another.

In the USA, there still remain a lot of circuits which are very similar to ISDN but not identical. These are called 'Switched 56' and carry data at 56 kbit s rather than 64 kbit s^{-1} (the remaining 8 kbit s^{-1} that makes up the total of 64 kbit s^{-1} is used for housekeeping data in Switched 56, whereas with ISDN the housekeeping data is transferred in a 'D' channel on top of the two 64 kbit data channels). This can create some problems when trying to link ISDN terminals, if there is a Switched 56 bridge somewhere in the way, requiring file transfer to take place at the lower rate.

For many applications, ISDN services are being superceded by ADSL (Asymmetric Digital Subscriber Line) technology, allowing higher data rates than offered by the normal pair of ISDN B channels to be offered to consumers and business over conventional telephone lines. The two technologies are somewhat different though, and ISDN may be considered superior for real-time applications requiring switched circuits and quality of service guarantees.

6.6.7 Protocols for the Internet

The Internet is now established as a universal means for worldwide communication. Although real-time protocols and quality of service do not sit easily with the idea of a free-for-all networking structure, there is growing evidence of applications that allow real-time audio and video information to be streamed with reasonable quality. The RealAudio format, for example, developed by Real Networks, is designed for coding audio in streaming media applications, currently at rates between 12 and 352 kbit s^{-1} for stereo audio, achieving respectable quality at the higher rates. People are also increasingly using the Internet for transferring multimedia projects between sites using FTP (file transfer protocol).

The Internet is a collection of interlinked networks with bridges and routers in various locations, which originally developed amongst the academic and research community. The bandwidth (data rate) available on the Internet varies from place to place, and depends on the route over which data is transferred. In this sense there is no easy way to guarantee a certain bandwidth, nor a certain 'time slot', and when there is a lot of traffic it simply takes a long time for data transfers to take place. Users access the Internet through a service provider (ISP), using either a telephone line and a modem, ISDN or an ADSL connection. The most intensive users will probably opt for high-speed leased lines giving direct access to the Internet.

The common protocol for communication on the Internet is called TCP/IP (Transmission Control Protocol/Internet Protocol). This provides a connection-oriented approach to data transfer, allowing for verification of packet integrity, packet order and retransmission in the case of packet loss. At a more detailed level, as part of the TCP/IP structure, there are high level protocols for transferring data in different ways. There is a file transfer protocol (FTP) used for downloading files from remote sites, a simple mail transfer protocol (SMTP) and a post office protocol (POP) for transferring email, and a hypertext transfer protocol (HTTP) used for interlinking sites on the world-wide web (WWW). The WWW is a collection of file servers connected to the Internet, each with its own unique IP address (the method by which devices connected to the Internet are identified), upon which may be stored text, graphics, sounds and other data.

UDP (user datagram protocol) is a relatively low-level connectionless protocol that is useful for streaming over the Internet. Being connectionless, it does not require any handshaking between transmitter and receiver, so the overheads are very low and packets can simply be streamed from a transmitter without worrying about whether or not the receiver gets them. If packets are missed by the receiver, or received in the wrong order, there is little to be done about it except mute or replay distorted audio, but UDP can be efficient when bandwidth is low and quality of service is not the primary issue.

Various real-time protocols have also been developed for use on the Internet, such as RTP (real-time transport protocol). Here packets are time-stamped and may be reassembled in the correct order and synchronised with a receiver clock. RTP does not guarantee quality of service or reserve bandwidth but this can be handled by a protocol known as RSVP (reservation protocol). RTSP is the real-time streaming protocol that manages more sophisticated functionality for streaming media servers and players, such a stream control (play, stop, fast-forward, etc.) and multicast (streaming to numerous receivers).

6.6.8 Wireless networks

Increasing use is made of wireless networks these days, the primary advantage being the lack of need for a physical connection between devices. There are various IEEE 802 standards for wireless networking, including 802.11 which covers wireless Ethernet or 'Wi-Fi'. These typically operate on either the 2.4 GHz or 5 GHz radio frequency bands, at relatively low power, and use various interference reduction and avoidance mechanisms to enable networks to coexist with other services. It should, however, be recognised that wireless networks will never be as reliable as wired networks owing to the differing conditions under which they operate, and that any critical applications in which real-time streaming is required would do well to stick to wired networks where the chances of experiencing drop-outs owing to interference or RF fading are almost non-existent. They are however extremely convenient for mobile applications and when people move around with computing devices, enabling reasonably high data rates to be achieved with the latest technology.

Bluetooth is one example of a wireless personal area network (WPAN) designed to operate over limited range at data rates of up to 1 Mbit s^{-1}. Within this there is the capacity for a number of channels of voice quality audio at data rates of 64 kbit s^{-1} and asynchronous channels up to 723 kbit s^{-1}. Taking into account the overhead for communication and error protection, the actual data rate achievable for audio communication is usually only sufficient to transfer data-reduced audio for a few channels at a time.

6.7 Streaming audio over computer interfaces

Desktop computers and consumer equipment are increasingly equipped with general purpose serial data interfaces such as USB (universal serial bus) and FireWire (IEEE 1394). These have a high enough data rate to carry a number of channels of audio data over relatively short distances, either over copper or optical fibre. Audio protocols exist for these, as described below. There are also a number of protocols designed to enable audio to be streamed in real time over general-purpose data networks such as Ethernet and ATM.

6.7.1 Audio over Firewire (IEEE 1394)

Firewire is an international standard serial data interface specified in IEEE 1394-1995. One of its key applications has been as a replacement for SCSI (Small Computer Systems Interface) for connecting disk drives and other peripherals to computers. It is extremely fast, running at rates of 100, 200 and 400 Mbit s^{-1} in its original form, with higher rates appearing all the time up to 3.2 Gbit s^{-1}. It is intended for optical fibre or copper interconnection, the copper 100 Mbit s^{-1} (S100) version being limited to 4.5 m between hops (a hop is the distance between two adjacent devices). The S100 version has a maximum realistic data capacity of 65 Mbit s^{-1}, a maximum of 16 hops between nodes and no more than 63 nodes on up to 1024 separate buses. On the copper version there are three twisted pairs – data, strobe and power – and the interface operates in half duplex mode, which means that communications in two directions are possible, but only one direction at a time. The 'direction' is determined by the current transmitter which will have arbitrated for access to the bus. Connections are 'hot pluggable' with auto-reconfiguration – in other words one can connect and disconnect devices without turning off the power and the remaining system will reconfigure itself accordingly. It is also relatively cheap to implement.

Unlike, for example, the AES3 audio interface, data and clock (strobe) signals are separated. A clock signal can be derived by exlusive-or'ing the data and strobe signals, as shown in Figure 6.25. Firewire combines features of network and point-to-point interfaces, offering both asynchronous and isochronous communication modes, so guaranteed latency and bandwidth are available if needed for time-critical applications. Communications are established between logical addresses, and the end point of an isochronous stream is called a 'plug'. Logical connections between devices can be specified as either 'broadcast' or 'point-to-point'. In the broadcast case either the transmitting or receiving plug is defined, but not both, and broadcast connections are unprotected in that any device can start and stop it. A primary advantage for audio applications is that point-to-point connections are protected – only the device that initiated a transfer can interfere with that connection, so once established the data rate is guaranteed for as long as the link remains intact. The interface can be used for real-time multichannel audio interconnections, file transfer, MIDI and machine control, carrying digital video, carrying any other computer data and connecting peripherals (e.g. disk drives).

Data is transferred in packets within a cycle of defined time (125 μs) as shown in Figure 6.26. The data is divided into 32 bit 'quadlets' and isochronous packets (which can be time stamped for synchronisation purposes) consist of between 1 and 256 quadlets (1024 bytes). Packet headers contain data from a cycle time register that allows for sample accurate timing to be

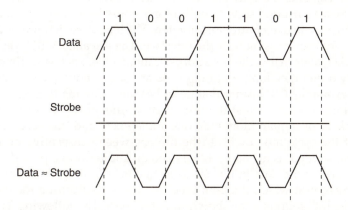

Figure 6.25 Data and strobe signals on the 1394 interface can be exclusive-or'ed to create a clock signal

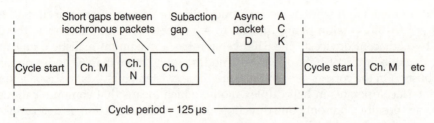

Figure 6.26 Typical arrangement of isochronous and asynchronous packets within a 1394 cycle

indicated. Resolutions down to about 40 nanoseconds can be indicated. One device on the bus acts as a bus master, initiating each cycle with a cycle start packet. Subsequently devices having isochronous packets to transmit do so, with short gaps between the packets, followed by a longer subaction gap after which any asynchronous information is transmitted.

Originating partly in Yamaha's 'm-LAN' protocol, the 1394 Audio and Music Data Transmission Protocol is now also available as an IEC PAS component of the IEC 61883 standard (a PAS is a publicly available specification that is not strictly defined as a standard but is made available for information purposes by organisations operating under given procedures). It offers a versatile means of transporting digital audio and MIDI control data. It specifies that devices operating this protocol should be capable of the 'arbitrated short bus reset' function which ensures that audio transfers are not interrupted during bus resets. Those wishing to implement this protocol should, of course, refer directly to the standard, but a short summary of some of the salient points is given here.

The complete model for packetising audio data so that it can be transported over the 1394 interface is complex and very hard to understand, but some applications make the overall structure seem more transparent, particularly if the audio samples are carried in a simple 'AM824' format, each quadlet of which has an 8-bit label and 24 bits of data. The model is layered as shown in Figure 6.27 in such a way that audio applications generate data that is formed (adapted) into blocks or clusters with appropriate labels and control information such as information about the nominal sampling frequency, channel configuration and so forth. Each block contains the information that arrives for transmission within one audio sample period, so in a surround sound application it could be a sample of data for each of six channels of audio plus related control information. The blocks, each representing 'events', are then 'packetised' for transmission over the interface. The so-called 'CIP layer' is the common isochronous packet layer that is the transport stream of 1394. Each isochronous packet has a header that is two quadlets long, defining it as an isochronous packet and indicating its length, and a two quadlet CIP header that describes the following data as audio/music data and indicates (among other things) the presentation time of the event for synchronisation purposes. A packet can contain more than one audio event and this becomes obvious when one notices that the cycle time of 1394 (the time between consecutive periods in which a packet can be transmitted) is normally 125 μs and an audio sample period at 48 kHz is only 22 μs. 1394 can carry audio data in IEC 60958 format (see Section 6.5.3). This is based on the AM824 data structure in which the 8-bit label serves as a substitute for the preamble and VUCP data of the IEC subframe, as shown in Figure 6.28. The following 24 bits of data are then simply the audio data component of the IEC subframe. The two subframes forming an

Example

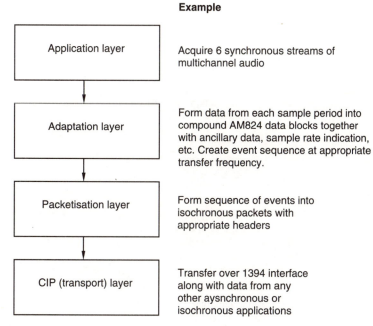

Acquire 6 synchronous streams of multichannel audio

Form data from each sample period into compound AM824 data blocks together with ancillary data, sample rate indication, etc. Create event sequence at appropriate transfer frequency.

Form sequence of events into isochronous packets with appropriate headers

Transfer over 1394 interface along with data from any other aysnchronous or isochronous applications

Figure 6.27 Example of layered model of 1394 audio/music protocol transfer

PAC = preamble code (takes place of preamble sync pattern in conventional digital interface)
11 = Z (or B)
01 = X (or M)
00 = Y (or W)

Figure 6.28 AM824 data structure for IEC 60958 audio data on 1394 interface. Other AM824 data types use a similar structure but the label values are different to that shown here

IEC frame are transmitted within the same event and each has to have the 8-bit label at the start of the relevant quadlet (indicating left or right channel).

The same AM824 structure can be used for carrying other forms of audio data including multibit linear audio (a raw audio data format used in some DVD applications, termed MBLA), high resolution MBLA, 1-bit audio (e.g. DSD), MIDI, SMPTE timecode and sample count or ancillary data. These are indicated by different 8-bit labels. One-bit audio can be either raw or DST (Direct Stream Transfer) encoded. DST is a lossless data reduction system employed in Direct Stream Digital equipment and Super Audio CD.

Audio data quadlets in these different modes can be clustered into compound data blocks. As a rule a compound data block contains samples from a number of related streams of audio

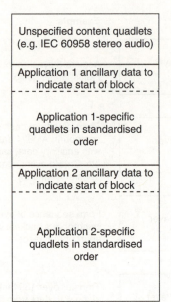

Figure 6.29 General structure of a compound data block

and ancillary information that are based on the same sampling frequency table (see Section 6.7.2). The parts of these blocks can be application specific or unspecific. In general, compound blocks begin with an unspecified region (although this is not mandatory) followed by one or more application-specific regions (see Figure 6.29). The unspecified region can contain audio/music content data and it is recommended that this always starts with basic two-channel stereo data in either IEC or raw audio format, followed by any other unspecified content data in a recommended order. An example of an application-specific part is the transfer of multiple synchronous channels from a DVD player. Here ancillary data quadlets indicate the starts of blocks and control factors such as downmix values, multichannel type (e.g. different surround modes), dynamic range control and channel assignment. An example of such a multichannel cluster is shown in Figure 6.30.

6.7.2 Audio over universal serial bus (USB)

The Universal Serial Bus is not the same as IEEE 1394, but it has some similar implications for desktop multimedia systems, including audio peripherals. USB has been jointly supported by a number of manufacturers including Microsoft, Digital, IBM, NEC, Intel and Compaq. Version 1.0 of the copper interface runs at a lower speed than 1394 (typically either 1.5 or 12 Mbit s^{-1}) and is designed to act as a low-cost connection for multiple input devices to computers such as joysticks, keyboards, scanners and so on. USB 2.0 runs at a higher rate up to 480 Mbit s^{-1} and is supposed to be backwards-compatible with 1.0.

USB 1.0 supports up to 127 devices for both isochronous and asynchronous communication and can carry data over distances of up to 5 m per hop (similar to 1394). A hub structure is

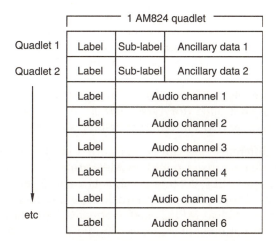

Quadlet 1

Quadlet 2

etc

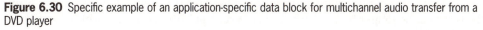

Figure 6.30 Specific example of an application-specific data block for multichannel audio transfer from a DVD player

required for multiple connections to the host connector. Like 1394 it is hot pluggable and reconfigures the addressing structure automatically, so when new devices are connected to a USB setup the host device assigns a unique address. Limited power is available over the interface and some devices are capable of being powered solely using this source – known as 'bus-powered' devices – which can be useful for field operation of, say, a simple A/D convertor with a laptop computer.

Data transmissions are grouped into frames of 1 ms duration in USB 1.0 but a 'micro-frame' of 1/8 of 1 ms was also defined in USB 2.0. A start-of-frame packet indicates the beginning of a cycle and the bus clock is normally at 1 kHz if such packets are transmitted every millisecond. So the USB frame rate is substantially slower than the typical audio sampling rate. The transport structure and different layers of the network protocol will not be described in detail as they are long and complex and can be found in the USB 2.0 specification. However it is important to be aware that transactions are set up between sources and destinations over so-called 'pipes' and that numerous 'interfaces' can be defined and run over a single USB cable, only dependent on the available bandwidth.

The way in which audio is handled on USB is well defined and somewhat more clearly explained than the 1394 audio/music protocol. It defines three types of communication: audio control, audio streaming and MIDI streaming. We are concerned primarily with audio streaming applications. Audio data transmissions fall into one of three types. Type 1 transmissions consist of channel-ordered PCM samples in consecutive sub-frames, while Type 2 transmissions typically contain non-PCM audio data that does not preserve a particular channel order in the bitstream, such as certain types of multichannel data-reduced audio stream. Type 3 transmissions are a hybrid of the two such that non-PCM data is packed into pseudo-stereo data words in order that clock recovery can be made easier. This method is in fact very much the same as the way data-reduced audio is packed into audio subframes within the IEC 61937 format described earlier in this chapter, and follows much the same rules.

Audio samples are transferred in subframes, each of which can be 1–4 bytes long (up to 24 bits resolution). An audio frame consists of one or more subframes, each of which represents a sample of different channel in the cluster (see below). As with 1394, a USB packet can contain a number of frames in succession, each containing a cluster of subframes. Frames are described by a format descriptor header that contains a number of bytes describing the audio data type, number of channels, subframe size, as well as information about the sampling frequency and the way it is controlled (for Type 1 data). An example of a simple audio frame would be one containing only two subframes of 24-bit resolution for stereo audio.

Audio of a number of different types can be transferred in Type 1 transmissions, including PCM audio (twos complement, fixed point), PCM-8 format (compatible with original 8-bit WAV, unsigned, fixed point), IEEE floating point, A-law and μ-law (companded audio corresponding to relatively old telephony standards). Type 2 transmissions typically contain data-reduced audio signals such as MPEG or AC-3 streams. Here the data stream contains an encoded representation of a number of channels of audio, formed into encoded audio frames that relate to a large number of original audio samples. An MPEG encoded frame, for example, will be typically be longer than a USB packet (a typical MPEG frame might be 8 or 24 ms long), so it is broken up into smaller packets for transmission over USB rather like the way it is streamed over the IEC 60958 interface described in Section 6.5.4. The primary rule is that no USB packet should contain data for more than one encoded audio frame, so a new encoded frame should always be started in a new packet. The format descriptor for Type 2 is similar to Type 1 except that it replaces subframe size and number of channels indication with maximum bit rate and number of audio samples per encoded frame. Currently only MPEG and AC-3 audio are defined for Type 2.

Rather like the compound data blocks possible in 1394 (see above), audio data for closely related synchronous channels can be clustered for USB transmission in Type 1 format. Up to 254 streams can be clustered and there are 12 defined spatial positions for reproduction, to simplify the relationship between channels and the loudspeaker locations to which they relate. (This is something of a simplification of the potentially complicated formatting of spatial audio signals and assumes that channels are tied to loudspeaker locations, but it is potentially useful. It is related to the channel ordering of samples within a WAVE format extensible file, described earlier.) The first six defined streams follow the internationally standardised order of surround sound channels for 5.1 surround, that is left, right, centre, LFE (low frequency enhancement), left surround, right surround. Subsequent streams are allocated to other loudspeaker locations around a notional listener. Not all the spatial location streams have to be present but they are supposed to be presented in the defined order. Clusters are defined in a descriptor field that includes 'bNrChannels' (specifying how many logical audio channels are present in the cluster) and 'wChannelConfig' (a bit field that indicates which spatial locations are present in the cluster). If the relevant bit is set then the relevant location is present in the cluster. The bit allocations are shown in Table 6.9.

6.7.3 AES 47: Audio over ATM

AES 47 defines a method by which linear PCM data, either conforming to AES 3 format or not, can be transferred over ATM. There are various arguments for doing this, not the least being the increasing use of ATM-based networks for data communications within the

Table 6.9 Channel identification in USB audio cluster descriptor

Data bit	Spatial location
D0	Left Front (L)
D1	Right Front (R)
D2	Center Front (C)
D3	Low Frequency Enhancement (LFE)
D4	Left Surround (Ls)
D5	Right Surround (Rs)
D6	Left of Center (Lc)
D7	Right of Center (Rc)
D8	Surround (S)
D9	Side Left (SL)
D10	Side Right (SR)
D11	Top (T)
D15..12	Reserved

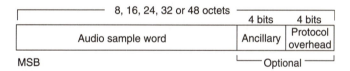

Figure 6.31 General audio subframe format of AES 47

broadcasting industry and the need to route audio signals over longer distances than possible using standard digital interfaces. There is also a need for low latency, guaranteed bandwidth and switched circuits, all of which are features of ATM. Essentially an ATM connection is established in a similar way to making a telephone call. A SETUP message is sent at the start of a new 'call' that describes the nature of the data to be transmitted and defines its vital statistics. The AES 47 standard describes a specific professional audio implementation of this procedure that includes information about the audio signal and the structure of audio frames in the SETUP at the beginning of the call.

For some reason bytes are termed octets in ATM terminology, so this section will follow that convention. Audio data is divided into subframes and each subframe contains a sample of audio as well as optional ancillary data and protocol overhead data, as shown in Figure 6.31. The setup message at the start of the call determines the audio mode and whether or not this additional data is present. The subframe should occupy a whole number of octets and the length of the audio sample should be such that the subframe is 8, 16, 24, 32 or 48 bits long. The ancillary data field, if it is present, is normally used for carrying the VUC bits from the AES 3 subframe, along with a B bit to replace the P (parity) bit of the AES 3 subframe (which has little relevance in this new application). The B bit in the '1' state indicates the start of an

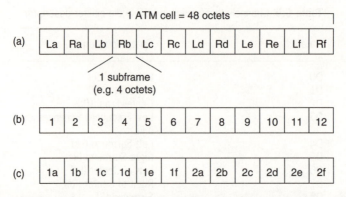

Figure 6.32 Packing of audio subframes into ATM cells. (a) Example of temporal ordering with two channels, left and right. 'a', 'b', 'c', etc., are successive samples in time for each channel. Co-temporal samples are grouped together. (b) Example of multichannel packing whereby concurrent samples from a number of channels are arranged sequentially. (c) Example of ordering by channel, with a number of samples from the same channel being grouped together. (If the number of channels is the same as the number of samples per cell, all three methods turn out to be identical.)

Table 6.10 Audio packing within ATM cells – options in AES 47

AAL code (hex)*	Subframe length (bytes)	Audio resolution	Ancillary bits	Protocol bits	Grouping	No. of audio channels
56 02	4	24	4	4	Temporal	2
56 01	4	24	4	4	N/A	1
06 02	3	24	0	0	Temporal	2
06 01	3	24	0	0	N/A	1
56 85	4	24	4	4	Multichannel	60

* This should be signalled within the second and third octets of the user-defined AAL part of the SETUP message that is an optional part of the ATM protocol for setting up calls between sources and destinations.

AES 3 channel status block, taking the place of the Z preamble that is no longer present. This data is transmitted in the order BCUV.

Samples are packed into the ATM cell either ordered in time, in multichannel groups or by channel, as shown in Figure 6.32. Only certain combinations of channels and data formats are allowed and all the channels within the stream have to have the same resolution and sampling frequency, as shown in Table 6.10.

Four octets in the user-defined AAL part of the SETUP message that begins a new ATM call define aspects of the audio communication that will take place. The first byte contains so-called 'qualifying information', only bit 4 of which is currently specified indicating that the sampling frequency is locked to some global reference. The second byte indicates the subframe format and sample length, while the third byte specifies the packing format. The fourth byte contains information about the audio sampling frequency (32, 44.1 or 48 kHz), its scaling factor (from 0.25 up to 8 times) and multiplication factor (e.g. $1/1.001$ or $1.001/1$ for 'pull-down' or 'pull-up'modes). It also has limited information for varispeed rates.

6.7.4 CobraNet

CobraNet is a proprietary audio networking technology developed by Peak Audio, a division of Cirrus Logic. It is designed for carrying audio over conventional Fast Ethernet networks (typically 100 Mbit s^{-1}), preferably using a dedicated Ethernet for audio purposes or using a switched Ethernet network. Switched Ethernet acts more like a telephone or ATM network where connections are established between specific sources and destinations, with no other data sharing that 'pipe'. For the reasons stated earlier in this chapter, Ethernet is not ideal for audio communications without some provisos being observed. CobraNet, however, implements a method of arbitration, bandwidth reservation and an isochronous transport protocol that enables it to be used successfully.

The CobraNet protocol has been allocated its own protocol identifier at the Data Link Layer of the ISO 7-layer network model, so it does not use Internet Protocol (IP) for data transport (this is typically inefficient for audio streaming purposes and involves too much overhead). Because it does not use IP it is not particularly suitable for wide area network (WAN) operation and would typically be operated over a local area network (LAN). It does however enable devices to be allocated IP addresses using the BOOTP (boot protocol) process and supports the use of IP and UDP (user datagram protocol) for other purposes than the carrying of audio. It is capable of transmitting packets in isochronous cycles, each packet transferring data for a 'bundle' of audio channels. Each bundle contains between 0 and 8 audio channels and these can either be unicast or multicast. Unicast bundles are intended for a single destination whereas multicast bundles are intended for 'broadcast' transmissions whereby a sending device broadcasts packets no matter whether any receiving device is contracted to receive them.

6.7.5 MAGIC

MAGIC (Media-accelerated Global Information Carrier) was developed by the Gibson Guitar Corporation, originally going under the name GMICS. It is a relatively recent audio interface that typically uses the Ethernet physical layer for transporting audio between devices, although it is not compatible with higher layers and does not appear to be interoperable with conventional Ethernet data networks. It uses its own application and data link layers, the data link layer of which is based on the Ethernet 802.3 data link layer, using a frame header that would be recognised by 802.3-compatible devices.

Although it is not limited to doing so, the described implementation uses 100 Mbit s^{-1} Fast Ethernet over standard CAT 5 cables, using four of the wires in a conventional Ethernet crossover implementation and the other four for power to devices capable of operating on limited power (9 volt, 500 mA). Data is formed into frames of 55 bytes, including relevant headers, and transmitted at a synchronous rate between devices. The frame rate is related to the audio sampling rate and a sampling clock can be recovered from the interface. Very low latency of 10–40 μs is claimed. MAGIC data can be daisy-chained between devices in a form more akin to point-to-point audio interfacing than computer networking, although routing and switching configurations are also possible using routing or switching hubs.

6.7.6 MOST

MOST (media oriented synchronous transfer) is an alternative network protocol designed for synchronous, asynchronous and control data over a low-cost optical fibre network. It is claimed that the technology sits between USB and IEEE 1394 in terms of performance and that MOST has certain advantages in the transfer of synchronous data produced by multimedia devices that are not well catered for in other protocols. It is stated that interfaces based on copper connections are prone to electromagnetic interference and that the optical fibre interface of this system provides immunity to such, in addition to allowing distances of up to 250 m between nodes in this case.

MOST specifies physical, data link and network layers in the OSI reference model for data networks and dedicated silicon has been developed for the physical layer. Data is transferred in 64-byteframes and the frame rate of data is dependent on the sampling rate in use by the connected devices, being 22.5 Mbit s^{-1} at a 44.1 kHz audio sampling rate. The bandwidth can be divided between synchronous and asynchronous data. Potential applications are described including professional audio, for transferring up to 15 stereo 16-bit audio channels or 10 stereo channels of 24-bit audio; consumer electronics, as an alternative to SPDIF at similar cost; automotive and home multimedia networking.

There is now a detailed specification framework for MOST (see Further reading) and it is the subject of a cooperation agreement between a number of manufacturers. It seems to have been most widely adopted in the automotive industry where it is close to being endorsed by a consortium of car makers.

6.6.7 BSS SoundWeb

BSS developed its own audio network interface for use with its SoundWeb products that are typically used in large venue installations and for live sound. It uses CAT 5 cabling over distances up to 300 m, but is not based on Ethernet and behaves more like a token ring network. Data is carried at a rate of about 12 Mbit s^{-1} and transports eight audio channels along with control information.

6.8 Digital content protection

Copy protection of digital content is increasingly required by the owners of intellectual property and data encryption is now regarded as the most appropriate way of securing such content from unwanted copying. The SCMS method used for copy protection on older interfaces such as IEC 60958 involved the use of two bits plus category codes to indicate the copy permission status of content, but no further attempt was made to make the audio content unreadable or to scramble it in the case of non-permitted transfers. A group of manufacturers known as 5C has now defined a method of digital content protection that is initially defined for IEEE 1394 transfers (see Section 6.7.1) but which is likely to be extended to other means of interconnection between equipment. It is written in a relatively generic sense, but the packet header descriptions currently refer directly to 1394 implementations. 5C is the five manufacturers Hitachi, Intel, Matsushita, Sony and Toshiba. The 1394 interface is

Table 6.11 Copy state indication in EMI bits of 1394 header

EMI bit states	Copy state	Authentication required
11	Copy never (Mode A)	Full
10	Copy one generation (Mode B)	Restricted or full
01	No more copies (Mode C)	Restricted or full
00	Copy free(ly) (Mode D)	None (not encrypted)

increasingly used on high-end consumer digital products for content transfer, although it has not been seen much on DVD and SACD players yet because the encryption model has only recently been agreed. There has also been the issue of content watermarking to resolve.

Content protection is managed in this model by means of both embedded copy control information (CCI) and by using two bits in the header of isochronous data packets (the so-called EMI or encryption mode indicator bits). Embedded CCI is that contained within the application-specific data stream itself. In other words it could be the SCMS bits in the channel status of IEC 60958 data or it could be the copy control information in an MPEG transport stream. This can only be accessed once a receiving device has decrypted the data that has been transmitted to it. In order that devices can inspect the copy status of a stream without decrypting the data, the packet header containing the EMI bits is not encrypted. Two EMI bits allow four copy states to be indicated as shown in Table 6.11.

The authentication requirement indicated by the copy state initiates a negotiation between the source and receiver that sets up an encrypted transfer using an exchanged key. The full details of this are beyond the scope of this book and require advanced understanding of cryptography, but it is sufficient to explain that full authentication involves more advanced cryptographic techniques than restricted authentication (which is intended for implementation on equipment with limited or reduced computational resources, or where copy protection is not a major concern). The negotiation process, if successful, results in an encrypted and decrypted transfer being possible between the two devices. Embedded CCI can then be accessed from within the content stream.

When there is a conflict between embedded CCI and EMI indications, as there might be during a stream (for example when different songs on a CD have different CCI but where the EMI setting remains constant throughout the stream) it is recommended that the EMI setting is the most strict of those that will be encountered in the transfer concerned. However the embedded CCI seems to have the final say-so when it comes to deciding whether the receiving device can record the stream. For example, even if EMI indicates 'copy never', the receiving device can still record it if the embedded CCI indicates that it is recordable. This ensures that a stream is as secure as it should be, and the transfer properly authenticated, before any decisions can be made by the receiving device about specific instances within the stream.

Certain AM824 audio applications (a specific form of 1394 Audio/Music Protocol interchange) have defined relationships between copy states and SCMS states, for easy translation when carrying data like IEC 60958 data over 1394. In this particular case the EMI 'copy never' state is not used and SCMS states are mapped onto the three remaining EMI states. For DVD applications the application-specific CCI is indicated in ancillary data and there is a mapping table specified for various relationships between this data and the indicated copy states. It depends to some extent on the quality of the transmitted data and whether or not it matches that indicated in the *audio_quality* field of ancillary data. (Typically DVD players have allowed single generation home copies of audio material over IEC 60958 interfaces at basic sampling rates, e.g. 48 kHz, but not at very high quality rates such as 96 kHz or 192 kHz.) SuperAudio CD applications currently have only one copy state defined and that is 'no more copies', presumably to avoid anyone duplicating the 1-bit stream that would have the same quality as the master recording.

Further reading

1394 Trade Association (2001) *TA Document 2001003: Audio and Music Data Transmission Protocol 2.0.*

5C (2001) *Digital transmission content protection specification, Volume 1 (informational version). Revision 1.2.* Available from: www.dtcp.com/.

AES (2002) *AES 47-2002: Transmission of digital audio over asynchronous transfer mode networks.*

Bailey, A. (2001) *Network Technology for Digital Audio.* Focal Press.

Gibson Guitar Corporation (2002) *Media-accelerated Global Information Carrier. Engineering Specification Version 2.4.* Available from: www.gibsonmagic.com.

IEC (1998) *IEC/PAS 61883-6. Consumer audio/video equipment – Digital interface – Part 6: Audio and music data transmission protocol.*

IEEE (1995) *IEEE 1394: Standard for a high performance serial bus.*

Oasis Technology (1999) *MOST Specification Framework v1.1.* Available from: www.oasis.com/technology/index.htm.

Page, M., Bentall, N., Cook, G. *et al.* (2002) Multichannel audio connection for Direct Stream Digital. Presented at *AES 113th Convention*, Los Angeles, Oct 5–8.

Philips (2002) *Direct Stream Digital Interchange File Format: DSD-IFF version 1.4, revision 2.* Available from: www.superaudiocd.philips.com.

Philips (2002) *Recommended usage of DSD-IFF, version 1.4.* Available from: www.superaudiocd.philips.com.

Rumsey, F. and Watkinson, J. (2004) *The Digital Interface Handbook*, 3rd edition. Focal Press.

USB (1998) *Universal serial bus: device class definition for audio devices, v1.0.*

Useful websites

Audio Engineering Society: www.aes.org
IEEE 1394: www.1394ta.org
Universal Serial Bus: www.usb.org

7 Audio software

This chapter provides a brief introduction to some of the most widely used software application categories, concentrating primarily on professional contexts as opposed to consumer ones. It does not attempt to explain how to use them in detail, as there are numerous other books that do that task very well, but it introduces the key concepts.

7.1 Sequencers

7.1.1 Introduction

Sequencers are probably the most ubiquitous of audio and MIDI applications. A sequencer will be capable of storing a number of 'tracks' of MIDI and audio information, editing it and otherwise manipulating it for musical composition purposes. It is also capable of storing MIDI events for non-musical purposes such as studio automation. Some of the more advanced packages are available in modular form (allowing the user to buy only the functional blocks required) and in cut-down or 'entry-level' versions for the new user. Popular packages such as ProTools and Logic now combine audio and MIDI manipulation in an almost seamless fashion, and have been developed to the point where they can no longer really be considered as simply sequencers. In fact they are full-blown audio production systems with digital mixers, synchronisation, automation, effects and optional video.

The dividing line between sequencer and music notation software is a grey one, since there are features common to both. Music notation software is designed to allow the user control over the detailed appearance of the printed musical page, rather as page layout packages work for typesetters, and such software often provides facilities for MIDI input and output. MIDI input is used for entering note pitches during setting, whilst output is used for playing the finished score in an audible form. Most major packages will read and write standard MIDI files, and can therefore exchange data with sequencers, allowing sequenced music to be exported to a notation package for fine tuning of printed appearance. It is also common for sequencer packages to offer varying degrees of music notation capability, although the scores that result may not be as professional in appearance as those produced by dedicated notation software.

7.1.2 Tracks, channels, instruments and environments

A sequencer can be presented to the user so that it emulates a multitrack tape recorder to some extent. The example shown in Figure 7.1 illustrates this point, showing the familiar transport controls as well as a multitrack 'tape-like' display.

A track can be either a MIDI track or an audio track, or it may be a virtual instrument of some sort, perhaps running on the same computer. A project is built up by successively overlaying more and more tracks, all of which may be replayed together. Tracks are not fixed in their time relationship and can be slipped against each other, as they simply consist of data stored in the memory. On older or less advanced sequencers, the replay of each MIDI track was assigned to a particular MIDI channel, but more recent packages offer an almost unlimited number of virtual tracks that can contain data for more than one channel (in order to drive a multitimbral instrument, for example). Using a multiport MIDI interface (see Chapter 5) it is

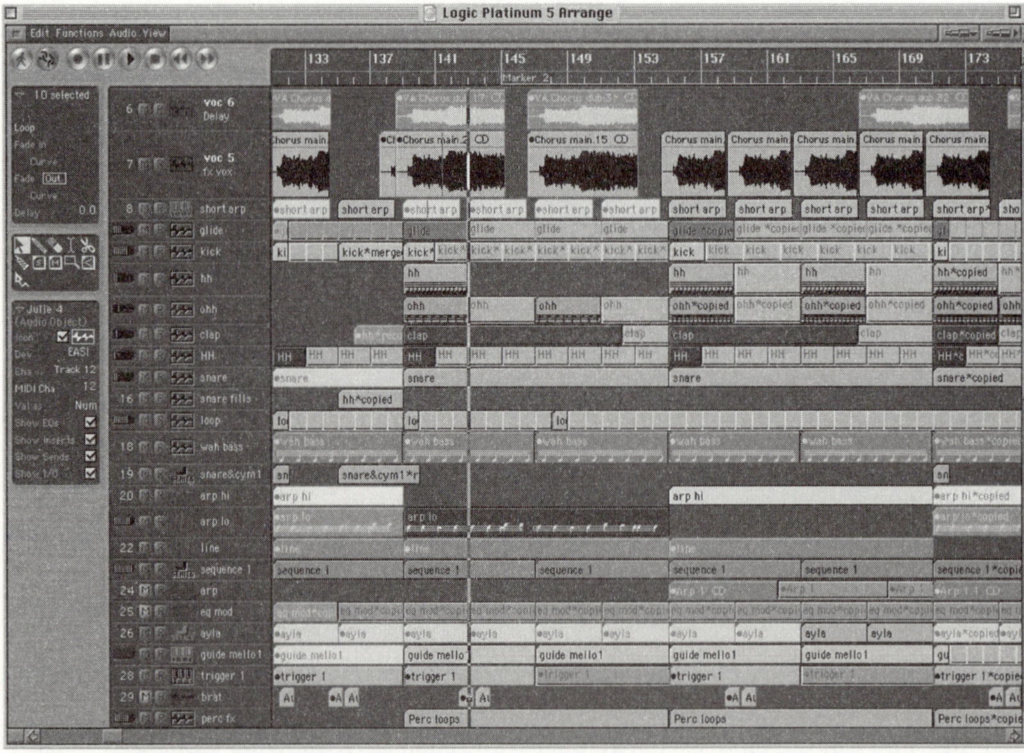

Figure 7.1 Example of a sequencer's primary display, showing tracks and transport controls. (Logic Platinum 5 'arrange' window)

possible to address a much larger number of instruments than the basic 16 MIDI channels allowed in the past.

In a typical sequencer, instruments are often defined in a separate 'environment' that defines the instruments, the ports to which they are connected, any additional MIDI processing to be applied, and so forth. An example is shown in Figure 7.2. When a track is recorded, therefore, the user simply selects the instrument to be used and the environment takes care of managing what that instrument actually *means* in terms of processing and routing. Now that soft synthesisers are used increasingly, sequencers can often address those directly via plug-in architectures such as DirectX or VST, without recourse to MIDI. These are often selected on pull-down menus for individual tracks, with voices selected in a similar way, often using named voice tables.

7.1.3 Input and output filters

As MIDI information is received from the hardware interface it will be stored in memory, but it may sometimes be helpful to filter out some information before it can be stored, using an input filter. This will be a sub-section of the program that watches out for the presence of certain MIDI status bytes and their associated data as they arrive, so that they can be discarded before storage. The user may be able to select input filters for such data as aftertouch,

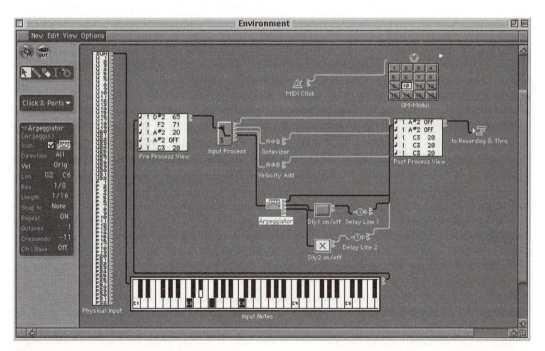

Figure 7.2 Example of environment window from Logic, showing ways in which various MIDI processes can be inserted between physical input and recording operation

pitch bend, control changes and velocity information, among others. Clearly it is only advisable to use input filters if it is envisaged that this data will never be needed, since although filtering saves memory space the information is lost for ever. Output filters are often implemented for similar groups of MIDI messages as for the input filters, acting on the replayed rather than recorded information. Filtering may help to reduce MIDI delays, owing to the reduced data flow.

7.1.4 Timing resolution

The timing resolution to which a sequencer can store MIDI events varies between systems. This 'record resolution' may vary with recent systems offering resolution to many thousandths of a note. Audio events are normally stored to sample accuracy. A sequencer with a MIDI resolution of 480 ppqn (pulses per quarter note) would resolve events to 4.1 millisecond steps, for example. The quoted resolution of sequencers, though, tends to be somewhat academic, depending on the operational circumstances, since there are many other factors influencing the time at which MIDI messages arrive and are stored. These include buffer delays and traffic jams. Modern sequencers have sophisticated routines to minimise the latency with which events are routed to MIDI outputs.

The record resolution of a sequencer is really nothing to do with the timing resolution available from MIDI clocks or timecode. The sequencer's timing resolution refers to the accuracy with which it time-stamps events and to which it can resolve events internally. Most sequencers attempt to interpolate or 'flywheel' between external timing bytes during replay, in an attempt to maintain a resolution in excess of the 24 ppqn implied by MIDI clocks (see Chapter 4).

7.1.5 Displaying, manipulating and editing information

A sequencer is the ideal tool for manipulating MIDI and audio information and this may be performed in a number of ways depending on the type of interface provided to the user. The most flexible is the graphical interface employed on many desktop computers which may provide for visual editing of the stored MIDI information either as a musical score, a table or event list of MIDI data, or in the form of a grid of some kind. Figure 7.3 shows a number of examples of different approaches to the display of stored MIDI information. Audio information is manipulated using an audio sample editor display that shows the waveform and allows various changes to be made to the signal, often including sophisticated signal processing, as discussed further below.

Although it might be imagined that the musical score would be the best way of visualising MIDI data, it is often not the most appropriate. This is partly because unless the input is successfully quantised (see below) the score will represent *precisely* what was played when the music was recorded and this is rarely good-looking on a score! The appearance is often messy because some notes were just slightly out of time. Score representation is useful after careful editing and quantisation, and can be used to produce a visually satisfactory printed output. Alternatively, the score can be saved as a MIDI file and exported to a music notation package for layout purposes.

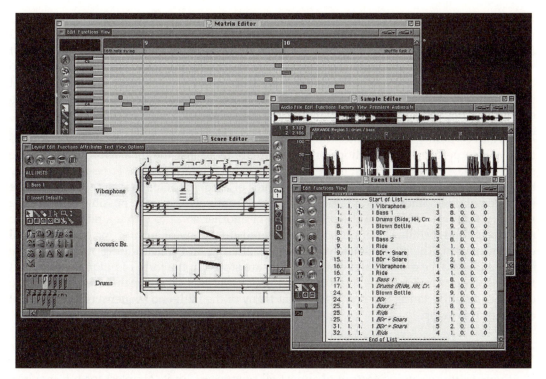

Figure 7.3 Examples of a selection of different editor displays from Logic, showing display of MIDI data as a score, a graphical matrix of events and a list of events. Audio can be shown as an audio waveform display

In the grid editing (called 'Matrix' in the example shown) display, MIDI notes may be dragged around using a mouse or trackball and audible feedback is often available as the note is dragged up and down, allowing the user to hear the pitch or sound as the position changes. Note lengths can be changed and the timing position may be altered by dragging the note left or right. In the event list form, each MIDI event is listed next to a time value. The information in the list may then be changed by typing in new times or new data values. Also events may be inserted and deleted. In all of these modes the familiar cut and paste techniques used in word processors and other software can be applied, allowing events to be used more than once in different places, repeated so many times over, and other such operations.

A whole range of semi-automatic editing functions are also possible, such as transposition of music, using the computer to operate on the data so as to modify it in a predetermined fashion before sending it out again. Echo effects can be created by duplicating a track and offsetting it by a certain amount, for example. Transposition of MIDI performances is simply a matter of raising or lowering the MIDI note numbers of every stored note by the relevant degree. Transposition of audio is more complicated, involving pitch-shifting algorithms. Recent pitch-shifting algorithms are so good, though, that they attempt to maintain the timbral quality of the voice while transposing it, in order that it does not end up sounding like 'Pinky and Perky' or 'Old Man River'. A number of algorithms have also been developed for converting audio melody lines to MIDI data, or using MIDI data to control the pitch of audio, further blurring the boundary between the two types of information. Silence can also be stripped from audio files, so that individual drum notes or vocal phrases can be turned into events in their own right, allowing them to be manipulated, transposed or time-quantised independently.

A sequencer's ability to search the stored data (both music and control) based on specific criteria, and to perform modifications or transformations to just the data which matches the search criteria, is one of the most powerful features of a modern system. For example, it may be possible to search for the highest-pitched notes of a polyphonic track so that they can be separated off to another track as a melody line. Alternatively it may be possible to apply the rhythm values of one track to the pitch values of another so as to create a new track, or to apply certain algorithmic manipulations to stored durations or pitches for compositional experimentation. The possibilities for searching, altering and transforming stored data are almost endless once musical and control events are stored in the form of unique values, and for those who specialise in advanced composing or experimental music these features will be of particular importance. It is in this field that many of the high-end sequencer packages will continue to develop.

7.1.6 Quantisation of rhythm

Rhythmic quantisation is a feature of almost all sequencers. In its simplest form it involves the 'pulling-in' of events to the nearest musical time interval at the resolution specified by the user, so that events that were 'out of time' can be played back 'in time'. It is normal to be able to program the quantising resolution to an accuracy of at least as small as a 32nd note and the choice depends on the audible effect desired. Events can be quantised either permanently or just for replay. Some systems allow 'record quantisation' which alters the timing of events as they arrive at the input to the sequencer. This is a form of permanent quantisation. It may also be possible to 'quantise' the cursor movement so that it can only drag events to predefined rhythmic divisions.

More complex rhythmic quantisation is also possible, in order to maintain the 'natural' feel of rhythm for example. Simple quantisation can result in music that sounds 'mechanical' and electronically produced, whereas the 'human feel' algorithms available in many packages attempt to quantise the rhythm strictly and then reapply some controlled randomness. The parameters of this process may be open to adjustment until the desired effect is achieved.

7.1.7 Automation and non-note MIDI events

In addition to note and audio events, one may either have recorded or may wish to add events for other MIDI control purposes such as program change messages, controller messages or system exclusive messages. Audio automation can also be added to control fades, panning, effects and other mixing features. Such data may be displayed in a number of ways, but again the graphical plot is arguably the most useful. It is common to allow automation data to be plotted as an overlay, such as shown in Figure 7.4.

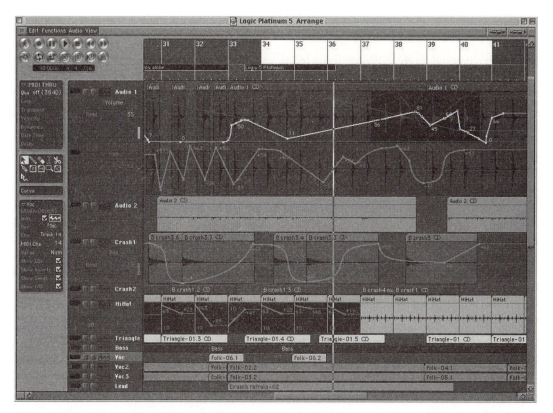

Figure 7.4 Example from Logic of automation data graphically overlaid on sequencer tracks

Some automation data is often stored in a so-called 'segment-based' form. Because automation usually relates to some form of audio processing or control, it usually applies to particular segments on the time-line of the current project. If the segment is moved, often one needs to carry the relevant automated audio processing with it. Segment-based processing or automation allows the changes in parameters that take place during a segment to be 'anchored' to that segment so that they can be made to move around with it if required.

It is possible to edit automation or control events in a similar way to note events, by dragging, drawing, adding and deleting points, but there are a number of other possibilities here. For example a scaling factor may be applied to controller data in order to change the overall effect by so many per cent, or a graphical contour may be drawn over the controller information to scale it according to the magnitude of the contour at any point. Such a contour could be used to introduce a gradual increase in MIDI note velocities over a section, or to introduce any other time-varying effect. Program changes can be inserted at any point in a sequence, usually either by inserting the message in the event list, or by drawing it at the appropriate point in the controller chart. This has the effect of switching the receiving device to a new voice or stored program at the point where the message is inserted. It can be used to ensure that all tracks in a sequence use the desired voices from the outset without having to set them up manually each time. Either the name of the program to be selected at that point

225

or its number can be displayed, depending on whether the sequencer is subscribing to a known set of voice names such as General MIDI.

System exclusive data may also be recorded or inserted into sequences in a similar way to the message types described above. Any such data received during recording will normally be stored and may be displayed in a list form. It is also possible to insert SysEx voice dumps into sequences in order that a device may be loaded with new parameters whilst a song is executing if required.

7.1.8 MIDI mixing and external control

Sequencers often combine a facility for mixing audio with one for controlling the volume and panning of MIDI sound generators. Using MIDI volume and pan controller numbers (decimal 7 and 10), a series of graphical faders can be used to control the audio output level of voices on each MIDI channel, and may be able to control the pan position of the source between the left and right outputs of the sound generator if it is a stereo source. On-screen faders may also be available to be assigned to other functions of the software, as a means of continuous graphical control over parameters such as tempo, or to vary certain MIDI continuous controllers in real time.

It is also possible with some packages to control many of the functions of the sequencer using external MIDI controllers. An external MIDI controller with a number of physical faders and buttons could be used as a basic means of mixing, for example, with each fader assigned to a different channel on the sequencer's mixer.

7.1.9 Synchronisation

A sequencer's synchronisation features are important when locking replay to external timing information such as MIDI clock or timecode. Most sequencers are able to operate in either beat clock or timecode sync modes and some can detect which type of clock data is being received and switch over automatically. To lock the sequencer to another sequencer or to a drum machine beat clock synchronisation may be adequate. If you will be using the sequencer for applications involving the timing of events in real rather than musical time, such as the dubbing of sounds to a film, then it is important that the sequencer is able to allow events to be tied to timecode locations, as timecode locations will remain in the same place even if the musical tempo is changed.

Sequencers incorporating audio tracks also need to be able to lock to sources of external audio or video sync information (e.g. word clock or composite video sync), in order that the sampling frequency of the system can be synchronised to that of other equipment in the studio.

7.1.10 Synchronised digital video

Digital video capability is now commonplace in desktop workstations. It is possible to store and replay full motion video on a desktop computer, either using a separate monitor or within a window on an existing monitor, using widely available technology such as QuickTime or Windows Multimedia Extensions. The replay of video from disk can be synchronised to the replay of audio and MIDI, using timecode, and this is particularly useful as an alternative to using video on a separate video tape recorder (which is mechanically much slower, especially in locating distant cues). In some sequencing or editing packages the video can simply be presented as another 'track' alongside audio and MIDI information.

In the applications considered here, compressed digital video is intended principally as a cue picture that can be used for writing music or dubbing sound to picture in post-production environments. In such cases the picture quality must be adequate to be able to see cue points and lip sync, but it does not need to be of professional broadcast quality. What is important is reasonably good slow motion and freeze-frame quality. Good quality digital video (DV), though, can now be transferred to and from workstations using a Firewire interface enabling video editing and audio post-production to be carried out in an integrated fashion, all on the one platform.

7.2 Plug-in architectures

7.2.1 What is a plug-in?

Plug-ins are now one of the fastest-moving areas of audio development, providing audio signal processing and effects that run either on a workstation's CPU or on dedicated DSP. (The hardware aspects of this were described in Chapter 5.) Audio data can be routed from a sequencer or other audio application, via an API (application programming interface) to another software module called a 'plug-in' that does something to the audio and then returns it to the source application. In this sense it is rather like inserting an effect into an audio signal path, but done in software rather than using physical patch cords and rack-mounted effects units. Plug-ins can be written for the host processor in a language such as C++, using the software development toolkits (SDK) provided by the relevant parties. Plug-in processing introduces a delay that depends on the amount of processing and the type of plug-in architecture used. Clearly low latency architectures are highly desirable for most applications.

Many plug-ins are versions of previously external audio devices that have been modelled in DSP, in order to bring favourite EQs or reverbs into the workstation environment. The sound quality of these depends on the quality of the software modelling that has been done. Some host-based (native) plug-ins do not have as good quality as dedicated DSP plug-ins as they may have been 'cut to fit' the processing power available, but as hosts become ever more powerful the quality of native plug-ins increases.

A number of proprietary architectures have been developed for plug-ins, including Microsoft's Direct X, Steinberg's VST, Digidesign's TDM, Mark of the Unicorn's MAS, TC Works' PowerCore and EMagic's host-based plug-in format. Apple's OS X Audio Units are a feature built in to the OS that manages plug-ins without the need for third-party middleware solutions. The popularity of this as a plug-in architecture has yet to be determined at the time of writing, but is likely to be used increasingly as OS X gains popularity. It is usually necessary to specify for which system any software plug-in is intended, as the architectures are not compatible. As OS-based plug-in architectures for audio become more widely used, the need for proprietary approaches may diminish.

Table 7.1 Digidesign plug-in alternatives

Plug-in architecture	Description
TDM	Uses dedicated DSP cards for signal processing. Does not affect the host CPU load and processing power can be expanded as required.
HTDM	(Host TDM). Uses the host processor for TDM plug-ins, instead of dedicated DSP.
RTAS	(Real Time Audio Suite). Uses host processor for plug-ins. Not as versatile as HTDM.
AudioSuite	Non-real-time processing that uses the host CPU to perform operations such as time-stretching that require the audio file to be rewritten.

Digidesign in fact has four different plug-in approaches that are used variously in its products, as shown in Table 7.1.

DirectX is a suite of multimedia extensions developed by Microsoft for the Windows platform. It includes an element called DirectShow that deals with real-time streaming of media data, together with the insertion of so-called 'filters' at different points. DirectX audio plug-ins work under DirectShow and are compatible with a wide range of Windows-based audio software. They operate at 32-bit resolution, using floating-point arithmetic and can run in real time or can render audio files in non-real time. They do not require dedicated signal processing hardware, running on the host CPU, and the number of concurrent plug-ins depends on CPU power and available memory. DirectX plug-ins are also scalable – in other words they can adapt to the processing resource available. They have the advantage of being compatible with the very wide range of DirectX-compatible software in the general computing marketplace but at the time of writing they can only handle two-channel audio.

DXi is a software synthesiser plug-in architecture developed by Cakewalk, running under DirectX. It is covered further in Section 7.3.

One example of a proprietary approach used quite widely is VST, Steinberg's Virtual Studio Technology plug-in architecture. It runs on multiple platforms and works in a similar way to DirectX plug-ins. On Windows machines it operates as a DLL (dynamic link library) resource, whereas on Macs it runs as a raw Code resource. It can also run on BeOS and SGI systems, as a Library function. VST incorporates both virtual effects and virtual instruments such as samplers and synthesisers. There is a cross-platform GUI development tool that enables the appearance of the user interface to be ported between platforms without the need to rewrite it each time.

7.2.2 Plug-in examples

Some examples of plug-in user interfaces are shown in Figure 7.5. An example of a stereo effects processor and a reverberation processor are shown. The quality of such plug-ins is now getting to the point where it is on a par with the sound quality achievable on external devices, depending primarily on the amount of DSP available.

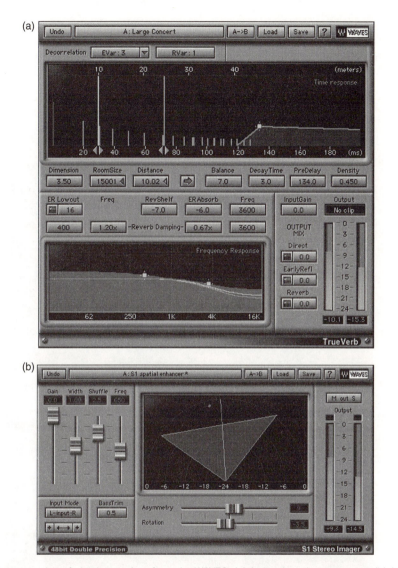

Figure 7.5 Two examples of plug-in user interfaces by WAVES. (a) A reverberation processor. (b) A stereo image processor

7.3 Virtual instruments

Virtual instruments or 'soft synths' and 'soft samplers' are software implementations of sound generators that can be controlled via the plug-in architecture. For example VSTi and DXi are examples of VST or DirectX virtual instruments. Mostly they rely on the host's CPU power to perform the synthesis operations and there is an increasing number of software versions of previously 'hard' sound generators. In many ways this can be quite convenient because it does away with the need for cumbersome external devices, MIDI cables and audio mixing. If all synthetic and sampled sound generation can be handled within the workstation, and the audio outputs of these virtual instruments mixed internally, the studio can really begin to be contained within a single box. External interfaces are then only required to handle acoustic sources such as vocals, guitars and other naturally recorded material.

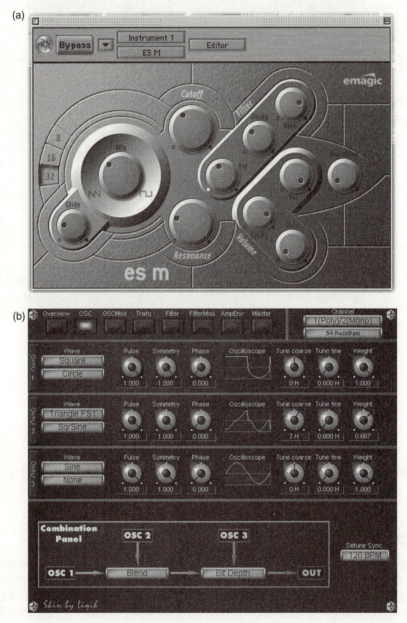

Figure 7.6 Two examples of plug-in synthesiser user interfaces. (a) Logic's ES synthesiser. (b) AnaMark VST synthesiser

Some examples of virtual instrument user interfaces are shown in Figure 7.6. They can usually be played either from MIDI tracks within a sequencer or by means of external MIDI controllers as stand-alone instruments.

7.4 Librarians and editors

Librarian and editor software is used for managing large amounts of voice data for MIDI-controlled instruments. As virtual instruments gradually take over from externally controlled devices the nature of librarians and editors will naturally evolve accordingly. Such packages communicate with external MIDI instruments using system exclusive messages in order to exchange parameters relating to voice programs. The software may then allow these voice programs or 'patches' to be modified using an editor, offering a rather better graphical interface than those usually found on the front panels of most sound modules. Banks of patches may be stored on disk by the librarian, in order that libraries of sounds can be managed easily, and this is often cheaper than storing patches in the various 'memory cards' offered by synth manufacturers. Banks of patch information can be accessed by sequencer software so that an operator can choose voices for particular tracks by name, rather than by program change numbers.

Sample editors are also available, offering similar facilities, although sample dumps using system exclusive are not really recommended, unless they are short, since the time taken can be excessive. Sample data can be transferred to a computer using a faster interface than MIDI (such as SCSI) and the sample waveforms can be edited graphically. A protocol known as SMDI ('smi-dee'), SCSI Musical Data Interchange, was developed by Matt Isaacson, based on the MIDI sample dump format, specifically for this purpose.

7.5 Audio editing and post-production software

Although most sequencers contain some form of audio editing these days, there are some software applications more specifically targeted at high quality audio editing and production. These have tended to come from a professional audio background rather than a MIDI sequencing background, although it is admitted that the two fields have met in the middle now and it is increasingly hard to distinguish a MIDI sequencer that has had audio tacked on from an audio editor that has had MIDI tacked on.

Audio applications such as those described here are used in contexts where MIDI is not particularly important and where fine control over editing crossfades, dithering, mixing, mastering and post-production functions are required. Here the editor needs tools for such things as: previewing and trimming edits, such as might be necessary in classical music post-production; PQ editing CD masters; preparing surround sound DVD material for encoding; MLP or AC-3 encoding of audio material; editing of DSD material for SuperAudio CD. The principles behind this technology were described in Chapters 2 and 3. The following two commercial examples demonstrate some of the practical concepts.

7.5.1 Sonic Studio HD

Sonic Studio HD is an example of a Mac-based product designed for preparing high quality CD, SACD and DVD audio material. It uses an approach to audio editing that relies on source and destination track windows, material being assembled by transferring selected takes from source to destination, with appropriate trims and crossfades. A typical user interface is shown in Figure 7.7. It is possible to see the audio waveform display and overlay of cross-fade information. Very detailed control is provided for the fine-tuning of crossfades, with numerous curve shapes and options, enabling the editor to modify edits until they sound right, as shown in Figure 7.8. Metering of multiple channels is provided in a separate window, as is listing and control of plug-ins for HDSP processing.

Like other systems of its kind, this product is capable of up-loading source material in the background so that takes to be edited can be copied from source tapes while one is editing in the foreground. It also provides comprehensive options for mastering, including PQ encoding for CDs and DDP list creation. Audio sampling rates up to 192 kHz are accommodated and options for editing DSD source material (see Chapter 2) are being developed (for Super Audio CD preparation). Audio processing is handled on a dedicated HDSP audio card, connected to an external audio interface.

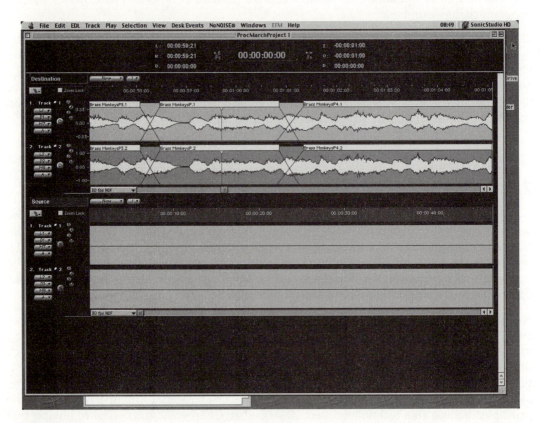

Figure 7.7 The main display of SonicStudio HD audio editor

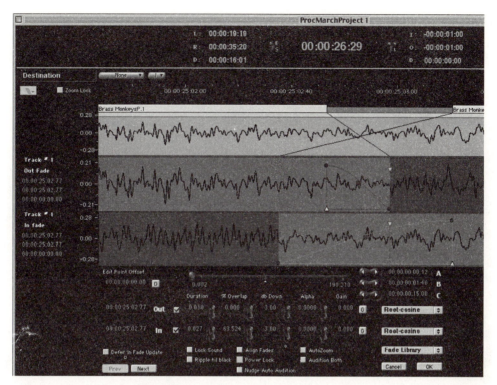

Figure 7.8 SonicStudio HD crossfade controls

7.5.2 SADiE

SADiE workstations run on the PC platform and most utilise an external audio interface. Recent Series 5 systems, however, can be constructed as an integrated rack-mounted unit containing audio interfaces and a Pentium PC. Both PCM and DSD signal processing options are available and the system makes provision for lossless MLP encoding for DVD-Audio, as well as SACD mastering and encoding.

A typical user interface for SADiE is shown in Figure 7.9. It is possible to see transport controls, the mixer interface and the playlist display. Audio can be arranged in the playlist by the normal processes of placing, dragging, copying and pasting, and there is a range of options for slipping material left or right in the list to accommodate new material (this ensures that all previous edits remain attached in the right way when the list is slipped backwards or forwards in time). Audio to be edited in detail can be viewed in the trim window (Figure 7.10) that shows a detailed waveform display, allowing edits to be previewed either to or from the edit point, or across the edit, using the play controls in the top right-hand corner (this is particularly useful for music editing). The crossfade region is clearly visible, with different colours and shadings used to indicate the 'live' audio streams before and after the edit.

The latest software allows for the use of DirectX plug-ins for audio processing. A range of plug-ins is available, including CEDAR audio restoration software, as described below.

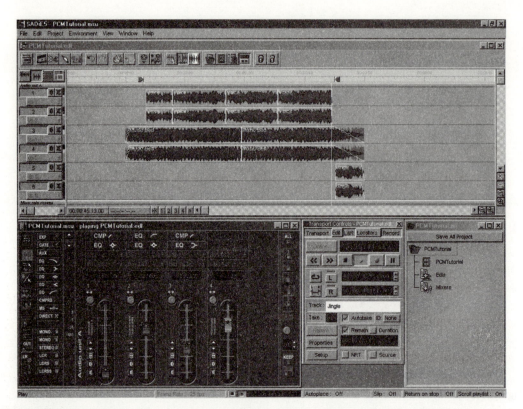

Figure 7.9 SADiE editor displays, showing mixer, playlist, transport controls and project elements

Figure 7.10 SADiE trim window showing crossfade controls and waveform display

7.6 Mastering and restoration software

Some software applications are designed specifically for the mastering and restoration markets. These products are designed either to enable 'fine tuning' of master recordings prior to commercial release, involving subtle compression, equalisation and gain adjustment (mastering), or to enable the 'cleaning up' of old recordings that have hiss, crackle and clicks (restoration).

CEDAR applications or plug-ins are good examples of the restoration group. Sophisticated controls are provided for the adjustment of de-hissing and de-crackling parameters, which often require considerable skill to master. Recently the company has introduced advanced visualisation tools that enable restoration engineers to 'touch up' audio material using an interface not dissimilar to those used for photo editing on computers. Audio anomalies (unwanted content) can be seen in the time and frequency domains, highlighted and interpolated based on information either side of the anomaly. A typical display from its RETOUCH product for the SADiE platform is shown in Figure 7.11.

CEDAR's restoration algorithms are typically divided into 'decrackle', 'declick', 'dethump' and 'denoise', each depending on the nature of the anomaly to be corrected. Some typical user interfaces for controlling these processes are shown in Figure 7.12.

Mastering software usually incorporates advanced dynamics control such as the TC Works Master X series, based on its Finalizer products, a user interface of which is pictured in Figure 7.13. Here compressor curves and frequency dependency of dynamics can be adjusted and metered. The display also allows the user to view the number of samples at peak level to watch for digital overloads that might be problematic.

7.7 Advanced audio processing software and development tools

High-end audio signal processing workstations, such as the Lake Huron, are designed primarily for research and development purposes. There is also a range of signal processing software for audio research and development that can run on general purpose desktop computers. Although this is not the primary emphasis of this book, brief mention will be made.

Signal processing workstations such as the Huron use large amounts of dedicated DSP hardware to enable the development of advanced real-time algorithms and signal analysis processes. Systems such as this are used for tasks such as acoustical modelling and real-time rendering of complex virtual reality scenes that require many hundreds of millions of computations per second. Such operations are typically beyond the scope of the average desktop PC, requiring some hours of off-line 'number crunching'. Using high-end workstations such processes may be run off-line in a fraction of the time or may be implemented in real time. A range of applications is available for the Huron workstation, ranging from head-tracked binaural simulation to virtual acoustic reality development tools. Interfaces are available between Huron and MATLAB, the latter being a popular research tool for the analysis, visualisation and manipulation of data.

MSP is a signal processing toolbox and development environment based on the Max MIDI programming environment described below. MSP runs on the Mac or SGI platforms and is designed to enable users to assemble signal processing 'engines' with a variety of components

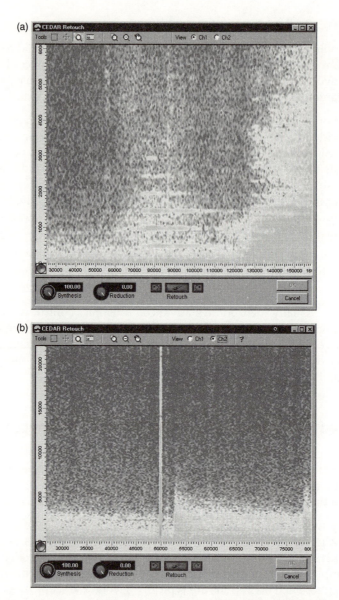

Figure 7.11 CEDAR Retouch display for SADiE, showing frequency (vertical) against time (horizontal) and amplitude (colour/density). Problem areas of the spectrographic display can be highlighted and a new signal synthesised using information from the surrounding region. (a) Harmonics of an interfering signal can be clearly seen. (b) A short-term spike crosses most of the frequency range

(either library or user-defined). They are linked graphically in a similar manner to the MIDI programming objects used in Max, allowing switches, gain, equalisation, delays and other signal processing devices to be inserted in the signal chain. For the user that is not conversant with programming DSPs directly, MSP provides an easy way in to audio signal processing, by pre-defining the building blocks and enabling them to be manipulated and linked graphically.

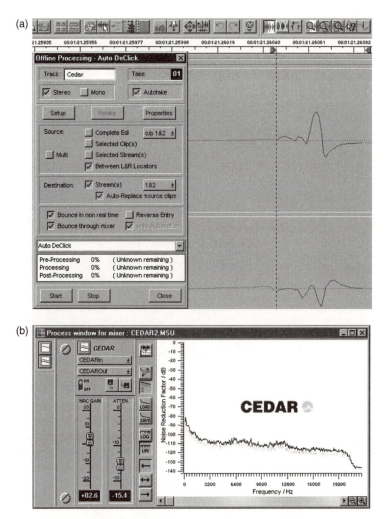

Figure 7.12 CEDAR restoration plug-ins for SADiE, showing (a) Declick and (b) Denoise processes

Signal processing can be run on the host CPU, provided it is sufficiently fast. An example of an MSP patch that acts as a variable stereo delay processor is shown in Figure 7.14.

7.8 Computer music software

There is a vast range of computer music software available for just about every platform, much of which is free or experimental or shareware. The computer music world is often strongly based in universities and research establishments, leading to a range of little known (outside these fields) but interesting applications that circulate mainly among the cognoscenti and may be downloadable free or for a reasonable price on the Internet. It is not the intention to review this field in detail here, as this book is concerned primarily with mainstream audio production tools. Eduardo Reck Miranda's book *Computer Sound Design: Synthesis Techniques and Programming* provides an excellent introduction to some aspects of

Figure 7.13 TC Works MasterX mastering dynamics plug-in interface

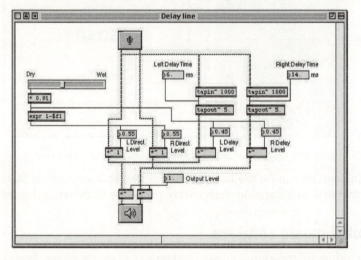

Figure 7.14 Example of a Max MSP patch that describes a variable stereo delay processor

this field, along with a CD-ROM containing a number of programs exemplifying the techniques described in the book.

A good example of a MIDI programming environment used widely in computer music is 'MAX', named after one of the fathers of electronic music, Max Matthews. MAX is essentially a MIDI programmer's construction kit, with numerous pre-defined functions. It is now

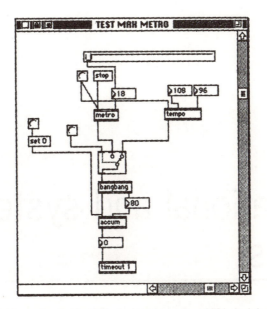

Figure 7.15 A simple 'patch' constructed in MAX, designed to output MIDI clock bytes at a rate determined by either the tempo or the metro objects

handled by Cycling '74 although it was originally supported by Opcode. It also allows new objects to be written in 'C' for those whose ambitions extend further than these. MAX allows functional objects to be linked graphically by dragging 'wires' from outputs to inputs, making it possible for the user to construct virtually any MIDI control 'engine' out of the building blocks. A worldwide network of MAX developers exists, and a large number of third party objects have been authored, many of which are available for the asking. An example of a MAX program or 'patch' is shown in Figure 7.15. This simple program sends MIDI clock bytes to a predefined output port at a rate determined by the 'tempo' or 'metro' objects.

Further reading

Collins, M. (2002) *ProTools 5.1 for Music Production*. Focal Press.
Miranda, E. R. (2002) *Computer Sound Design: Synthesis Techniques and Programming*. Focal Press.

8 Operational and systems issues

This chapter is concerned with technical, operational and systems issues that may arise when using computer-based audio hardware and software. It is not intended as a 'how to' chapter, because all software packages are different and other books are available that explain them, but it does deal with some of the more specialised issues that may arise in high quality professional audio environments and familiarises the reader with preparations for various consumer release formats and media.

8.1 Level control and metering

Typical audio systems today have a very wide dynamic range that equals or exceeds that of the human hearing system. Distortion and noise inherent in the recording or processing of audio are at exceptionally low levels owing to the use of high resolution A/D convertors, up to 24-bit storage, and wide range floating-point signal processing. This is not to say that quality is perfect. It is more intended to support the assertion that level control is less crucial than it used to be in the days when a recording engineer struggled to optimise a recording's dynamic range between the noise floor and the distortion ceiling (see Figure 8.1).

The dynamic range of a typical digital audio system can now be well over 100 dB and there is room for the operator to allow a reasonable degree of 'headroom' between the peak audio signal level and the maximum allowable level. Meters are provided to enable the signal level to be observed, and they are usually calibrated in dB, with zero at the top and negative dBs below this. The full dynamic range is not always shown, and there may be a peak bar that can hold the maximum level permanently or temporarily. As explained in Chapter 2, 0 dBFS (full scale) is the point at which all of the bits available to represent the signal have been used. Above this level the signal clips and the effect of this is quite objectionable, except on very short transients where it may not be noticed. It follows that signals should never be allowed to clip.

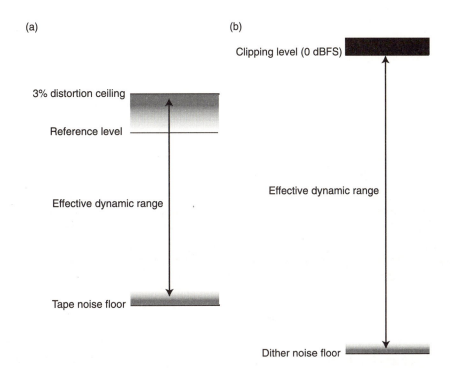

Figure 8.1 Comparison of analog and digital dynamic range. (a) Analog tape has increasing distortion as the recording level increases, with an effective maximum output level at 3% third harmonic distortion. (b) Modern high-resolution digital systems have wider dynamic range with a noise floor fixed by dither noise and a maximum recording level at which clipping occurs. The linearity of digital systems does not normally become poorer as signal level increases, until 0 dBFS is reached. This makes level control a somewhat less important issue at the initial recording stage, provided sufficient headroom is allowed for peaks

There is a tendency in modern audio production to want to master everything so that it sounds as loud as possible, and to ensure that the signal peaks as close to 0 dBFS as possible. This level maximising or normalising process can be done automatically in most packages, the software searching the audio track for its highest level sample and then adjusting the overall gain so that this just reaches 0 dBFS. In this way the recording can be made to use all the bits available, which can be useful if it is to be released on a relatively low-resolution consumer medium where noise might be more of a problem. (It is important to make sure that correct redithering is used when altering the level and requantising, as explained in Section 2.8.) This does not, of course, take into account any production decisions that might be involved in adjusting the overall levels of individual tracks on an album or other compilation, where relative levels should be adjusted according to the nature of the individual items, their loudness and the producer's intent.

A little-known but important fact is that even if the signal is maximised in the automatic fashion, so that the highest sample value just does not clip, subsequent analog electronics in the signal chain may still do so. Some equipment is designed in such a way that the maximum digital signal level is aligned to coincide with the clipping voltage of the analog electronics in a D/A convertor. In fact, owing to the response of the reconstruction filter in the D/A convertor (which reconstructs an analog waveform from the PAM pulse train) inter-sample signal peaks can be created that slightly exceed the analog level corresponding to 0 dBFS, thereby clipping the analog side of the convertor. For this reason it is recommended that digital-side signals are maximised so that they peak a few dB below 0 dBFS, in order to avoid the distortion that might otherwise result on the analog side. Some mastering software provides detailed analysis of the signal showing exactly how many samples occur in sequence at peak level, which can be a useful warning of potential or previous clipping.

8.2 Spatial reproduction formats

Now that two-channel stereo is no longer the ubiquitous release format that it was, users need to understand something about alternative multichannel reproduction formats, such as used for surround sound. These are used widely on DVD, SACD and for television, games and movie production. Audio applications now provide many facilities for working in these multichannel formats. This section is a short introduction, more detailed explanation of which can be found in my book *Spatial Audio*.

8.2.1 Introduction to multichannel formats

Whereas two-channel stereo normally employs two loudspeakers at ±30° in front of the listener, creating a stereophonic 'scene' or 'panorama' between the loudspeakers, multichannel stereo or surround sound employs more loudspeakers to increase the sense of realism and spatial complexity. A variety of formats have been tried over the years, including things like quadraphonic sound in the 1970s, but the most common production formats at the time of writing are formats based on movie-style surround sound. These typically employ a number of front channels to enable accurate phantom imaging and a number of side or rear channels that are used for ambience or effects. The centre channel (not present in two-channel stereo) has the effect of widening the listening area and anchoring dialogue or vocals in the centre of the image, even for off-centre listeners.

In international standards describing stereo loudspeaker configurations the nomenclature for the configuration is often in the form '*n-m stereo*', where *n* is the number of front channels and *m* is the number of rear or side channels. This distinction can be helpful as it reinforces the slightly different role of the surround channels, although many simply refer to these formats as *x*-channel surround, making no distinction between front and rear channels.

Audio in these multichannel formats is often encoded for consumer release using low bit-rate coding algorithms such as Dolby Digital, although it can be stored in linear PCM form if sufficient space or bandwidth is available. Examples of these coding approaches were given in Section 2.12.

8.2.2 4-channel surround (3-1 stereo)

'3-1 stereo', or 'LCRS surround', is a format used quite widely in analog cinema installations and older home cinema systems that used Dolby matrix encoding and decoding technology. In the 3-1 approach a single 'effects' or 'surround' channel is routed to a loudspeaker or loudspeakers located behind (and possibly to the sides) of listeners. It was developed first for cinema applications, enabling a greater degree of audience involvement in the viewing/listening experience by providing a channel for 'wrap-around' effects. There is no specific intention in 3-1 stereo to use the effects channel as a means of enabling 360° image localisation. In any case, this would be virtually impossible with most configurations as there is only a single audio channel feeding a larger number of surround loudspeakers, effectively in mono.

Figure 8.2 shows the typical loudspeaker configuration for this format. In the cinema there are usually a large number of surround loudspeakers fed from the single channel, in order to cover a wide audience area. This has the tendency to create a relatively diffuse or distributed reproduction of the effects signal. The surround speakers are sometimes electronically decorrelated to increase the degree of spaciousness or diffuseness of surround effects, in order that they are not specifically localised to the nearest loudspeaker or perceived inside the head. In consumer systems reproducing 3-1 stereo, the mono surround channel is normally fed to two surround loudspeakers located in similar positions to the 3-2 format described below (these are dipoles in the Home THX system, so as to create a more diffuse spatial effect). The gain

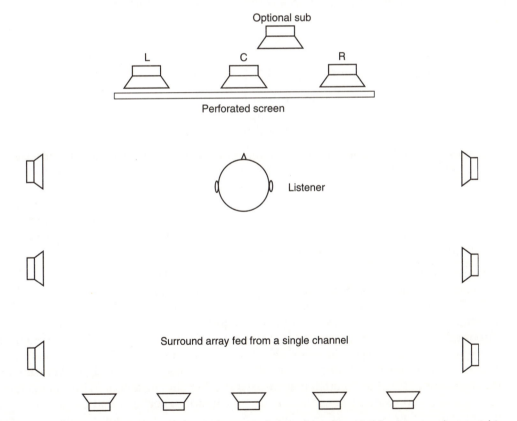

Figure 8.2 3-1 format reproduction uses a single surround channel usually routed (in cinema environments) to an array of loudspeakers to the sides and rear of the listening area. In consumer reproduction the mono surround channel may be reproduced through only two surround loudspeakers, possibly using artificial decorrelation and/or dipole loudspeakers to emulate the more diffused cinema experience

of the channel is usually reduced by 3 dB so that the summation of signals from the two speakers does not lead to a level mismatch between front and rear.

This surround format is usually matrix encoded using a Dolby Stereo encoder. This takes the four channels and combines them in a manner that creates a two-channel compatible signal from which the centre and surround information can subsequently be extracted if required. The reason for this was compatibility with two-channel analog media and to enable surround audio to be encoded on two optical film sound tracks. Matrix decoding often involves some sort of active 'steering' to increase the channel separation, such as employed in Dolby Prologic decoders. Dolby Stereo or Surround is normally monitored and mixed through an encoder and decoder matrix, in order to hear the effect, as it is not a perfect process.

The mono surround channel is the main limitation in this format. Despite the use of multiple loudspeakers to reproduce the surround channel, it is still not possible to create a good sense of envelopment or spaciousness without using surround signals that are different on both sides of the listener. Most of the psychoacoustics research suggests that the ears need to be provided with decorrelated signals to create the best sense of envelopment and effects can be better spatialised using stereo surround channels.

8.2.3 5.1 channel surround (3-2 stereo)

The 3-2 configuration has been standardised for numerous surround sound applications, including cinema, television and consumer applications. Because of its wide use in general parlance, though, the term '5.1 surround' will be used.

The mono surround limitation is removed in the 5.1-channel system, enabling the provision of stereo effects or room ambience to accompany a primarily front-oriented sound stage. Essentially the front three channels are intended to be used for a conventional three-channel stereo sound image, while the rear/side channels are only intended for generating supporting ambience, effects or 'room impression'. In this sense, the standard does not directly support the concept of 360° image localisation, although it may be possible to arrive at recording techniques or signal processing methods that achieve this to a degree.

One cannot introduce the 5.1 surround system without explaining the meaning of the '.1' component. This is a dedicated low-frequency effects (LFE) channel or sub-bass channel. It is called '.1' because of its limited bandwidth (normally up to 120 Hz). It is intended for conveying special low-frequency content that requires greater sound pressure levels and headroom than can be handled by the main channels. It is not intended for conveying the low-frequency component of the main channel signals, and its application is likely to be primarily in sound-for-picture applications where explosions and other high-level rumbling noises are commonplace, although it may be used in other circumstances. With cinema reproduction the in-band gain of this channel is usually 10 dB higher than that of the other individual channels. This is achieved by a level increase of the reproduction channel, not by an increased recording level. (This does not mean that the broadband or weighted SPL of the LFE loudspeaker should measure 10 dB higher than any of the other channels – in fact it will be considerably less than this as its bandwidth is narrower.)

The loudspeaker layout and channel configuration is specified in the ITU-R BS.775 standard. This is shown in Figure 8.3. A display screen is also shown in this diagram for sound with picture applications, and there are recommendations concerning the relative size of the screen and the loudspeaker base width shown in the standard. The left and right loudspeakers are located at $\pm30°$ for compatibility with two channel stereo reproduction. In many ways this need for compatibility with 2/0 is a pity, because the centre channel unavoidably narrows the front sound stage in many applications, and the front stage could otherwise take

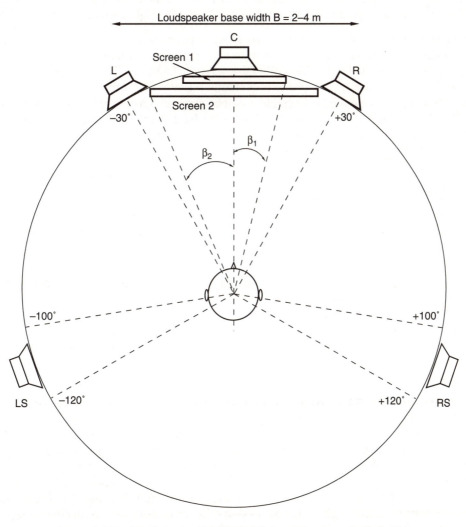

Screen 1: Listening distance = 3H ($2\beta_1 = 33°$) (possibly more suitable for TV screen)
Screen 2: Listening distance = 2H ($2\beta_2 = 48°$) (more suitable for projection screen)
H: Screen height

Figure 8.3 3-2 format reproduction according to the ITU-R BS.775 standard uses two independent surround channels routed to one or more loudspeakers per channel

advantage of the wider spacing facilitated by three-channel reproduction. It was nonetheless considered crucial for the same loudspeaker configuration to be usable for all standard forms of stereo reproduction, for reasons most people will appreciate.

In the 5.1 standard there are normally no loudspeakers directly behind the listener, which can make for creative difficulties. This has led to a Dolby proposal called EX (described below) that places an additional speaker at the centre-rear location. (This is not part of the current standard, though.) The ITU standard also allows for additional surround loudspeakers to cover the region around listeners, similar to the 3-1 arrangement described earlier. If these are used then they are expected to be distributed evenly in the angle between ±60° and ±150°.

The limitations of the 5.1 format, particularly in some peoples' view for music purposes, have led to various non-standard uses of the five or six channels available on new consumer disk formats such as DVD-A (Digital Versatile Disk – Audio) and SACD (Super Audio Compact Disc). For example, some are using the sixth channel (that would otherwise be LFE) in its full bandwidth form on these media to create a height channel. Others are making a pair out of the 'LFE' channel and the centre channel so as to feed a pair of front-side loudspeakers, enabling the rear loudspeakers to be further back. These are non-standard uses and should be clearly indicated on any recordings.

8.2.4 Dolby EX

In 1998 Dolby and Lucasfilm THX joined forces to promote an enhanced surround system that added a centre rear channel to the standard 5.1-channel setup. They introduced it,

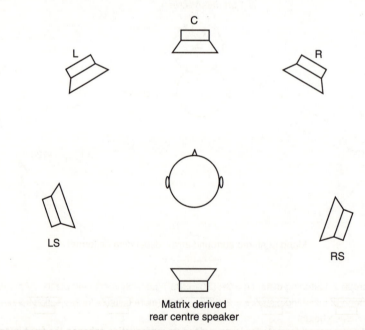

Figure 8.4 Dolby EX adds a centre-rear channel fed from a matrix-decoded signal that was originally encoded between left and right surround channels in a manner similar to the conventional Dolby Stereo matrix process

apparently, because of frustrations felt by sound designers for movies in not being able to pan sounds properly to the rear of the listener – the surround effect typically being rather diffuse. This system was christened 'Dolby Digital – Surround EX', and apparently uses matrix-style centre channel encoding and decoding between the left and right surround channels of a 5.1-channel mix. The loudspeakers at the rear of auditorium are then driven separately from those on the left and right sides, using the feed from this new 'rear centre' channel, as shown in Figure 8.4.

8.2.5 7.1 channel surround

Deriving from widescreen cinema formats, the 7.1 channel configuration normally adds two further loudspeakers to the 5.1 channel configuration, located at centre left (CL) and centre right (CR), as shown in Figure 8.5. This is not a format primarily intended for consumer applications, but for large cinema auditoria where the screen width is such that the additional channels are needed to cover the angles between the loudspeakers satisfactorily for all the seats in the auditorium. Sony's SDDS cinema system is the most common proprietary implementation of this format.

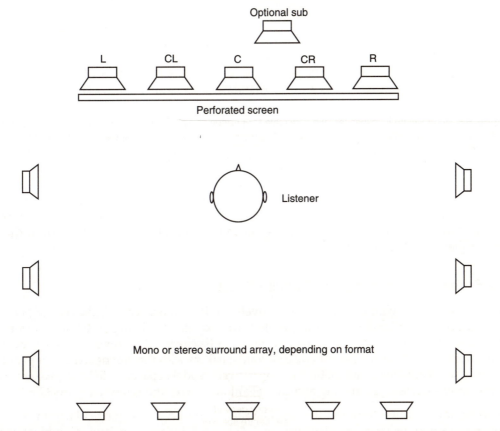

Figure 8.5 Some cinema sound formats for large auditorium reproduction enhance the front imaging accuracy by the addition of two further loudspeakers, centre left and centre right

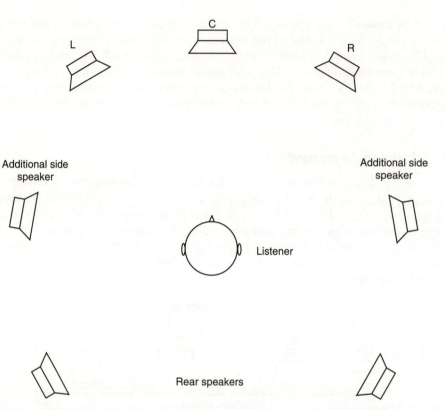

Figure 8.6 Approximate loudspeaker layout suitable for Lexicon's Logic 7 reproduction. Notice the additional side loudspeakers that enable a more enveloping image and may enable rear loudspeakers to be placed further to the rear

Lexicon and Meridian have also implemented a 7-channel mode in their consumer surround decoders, but the recommended locations for the loudspeakers are not quite the same as in the cinema application. The additional channels are used to provide a wider side-front component and allow the rear speakers to be moved round more to the rear than in the 5.1 arrangement (see Figure 8.6).

8.2.6 Surround panning and spatial effects

Pairwise amplitude panning, although relatively crude in many ways, is the type of pan control most commonly implemented in simple surround panners, being based on an extension of the two-channel sine/cosine panner to more loudspeakers. It involves adjusting the relative amplitudes between a pair of adjacent loudspeakers with the aim of creating a phantom image at some point between them. Panning between widely spaced side loudspeakers is not particularly successful at creating accurate phantom images though (see Figure 8.7).

In some applications designed for five-channel surround work, particularly in the film domain, separate panners are provided for L-C-R, LS-RS, and front–back. Combinations of positions of these amplitude panners enable sounds to be moved to various locations, some

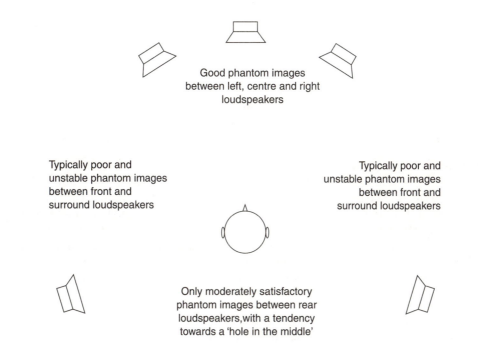

Good phantom images
between left, centre and right
loudspeakers

Typically poor and
unstable phantom images
between front and
surround loudspeakers

Typically poor and
unstable phantom images
between front and
surround loudspeakers

Only moderately satisfactory
phantom images between rear
loudspeakers,with a tendency
towards a 'hole in the middle'

Figure 8.7 Imaging accuracy in five-channel surround sound reproduction

more successfully than others. For example sounds panned so that some energy is emanating from all loudspeakers (say, panned centrally on all three pots) tend to sound diffuse for centre listeners and in the nearest loudspeaker for those sitting off centre. Joystick panners combine these amplitude relationships under the control of a single 'lever'. Moving effects made possible by these joysticks are often unconvincing and need to be used with experience and care.

Other more sophisticated panners may involve psychoacoustic filtering, binaural information or Ambisonic priniciples, and it is possible to encounter advanced spatial audio processing plug-ins that can be used to manipulate stereo images and alter spatial characteristics of implied environments. Distance and movement can be simulated effectively by changing direct/reverberant ratio, level, high frequency content and reflections, as well as Doppler shifts. These sometimes go hand in hand with reverberation processing, as this is one way of adding spatial content to a mix.

8.3 Controlling and maintaining sound quality

The sound quality achievable with modern workstations is now exceptionally high. As mentioned earlier in this chapter, there are now few technical reasons why distortion, noise, frequency response and other performance characteristics should not match the limits of human perception. Of course there will always be those for whom improvements can be made, but technical performance of digital audio systems is no longer really a major issue today.

If one accepts the foregoing argument, the maintenance of sound quality in computer-based production comes down more to understanding the operational areas in which quality can be compromised. These include things like ensuring as few A/D and D/A conversions as possible, maintaining audio resolution at 24 bits or more throughout the signal chain (assuming this is possible), redithering appropriately at points where requantising is done, and avoiding sampling frequency conversions. The rule of thumb should be to use the highest sampling frequency and resolution that one can afford to use, but no higher than strictly necessary for the purpose, otherwise storage space and signal processing power will be squandered. The scientific merits of exceptionally high sampling frequencies are dubious, for all but a few afficionados, although the marketing value may be considerable.

The point at which quality can be affected in a digital audio system is at A/D and D/A conversion. In fact the quality of an analog signal is irretrievably fixed at the point of A/D conversion, so this should be done with the best equipment available. There is very little that can be done afterwards to improve the quality of a poorly converted signal. At conversion stages the stability of timing of the sampling clock is crucial, because if it is unstable the audio signal will contain modulation artefacts that give rise to increased distortions and noise of various kinds. This so-called clock jitter is one of the biggest factors affecting sound quality in convertors and high quality external convertors usually have much lower jitter than the internal convertors used on PC sound cards.

The quality of a digital audio signal, provided it stays in the digital domain, is not altered unless the values of the samples are altered. It follows that if a signal is recorded, replayed, transferred or copied without altering sample values then the quality will not have been affected, despite what anyone may say. Sound quality, once in the digital domain, therefore depends entirely on the signal processing algorithms used to modify the program. There is little a user can do about this except choose high-quality plug-ins and other software, written by manufacturers that have a good reputation for DSP that takes care of rounding errors, truncation, phase errors and all the other nasties that can arise in signal processing. This is really no different from the problems of choosing good-sounding analog equipment. Certainly not all digital equaliser plug-ins sound the same, for example, because this depends on the filter design. Storage of digital data, on the other hand, does not affect sound quality at all, provided that no errors arise and that the signal is stored at full resolution in its raw PCM form (in other words, not MPEG encoded or some other form of lossy coding).

The sound quality the user hears when listening to the output of a workstation is not necessarily what the consumer will hear when the resulting program is issued on the release medium. One reason for this is that the sound quality depends on the quality of the D/A convertors used for monitoring. The consumer may hear better or worse, depending on the convertors used, assuming the bit stream is delivered without modification. One hopes that the convertors used in professional environments are better than those used by consumers, but this is not always the case. High-resolution audio may be mastered at a lower resolution for consumer release (e.g. 96 kHz, 24 bit recordings reduced to 44.1 kHz, 16 bits for release on CD), and this can affect sound quality. It is very important that any down-conversion of master recordings be done using the best dithering and/or sampling frequency conversion possible, especially when sampling frequency conversion is of a non-integer ratio.

Low bit-rate coders (e.g. MPEG) can reduce quality in the consumer delivery chain, but it is the content-provider's responsibility to optimise quality depending on the intended release format. Where there are multiple release formats it may be necessary to master the program differently in each case. For example, really low bit rate Internet streaming may require some enhancement (e.g. compression and equalisation) of the audio to make it sound reasonable under such unfavourable conditions.

When considering the authoring of interactive media such as games or virtual reality audio, there is a greater likelihood that the engineer, author, programmer and producer will have less control over the ultimate sound quality of what the consumer hears. This is because much of the sound material may be represented in the form of encoded 'objects' that will be rendered at the replay stage, as shown in Figure 8.8. Here the quality depends more on the quality of

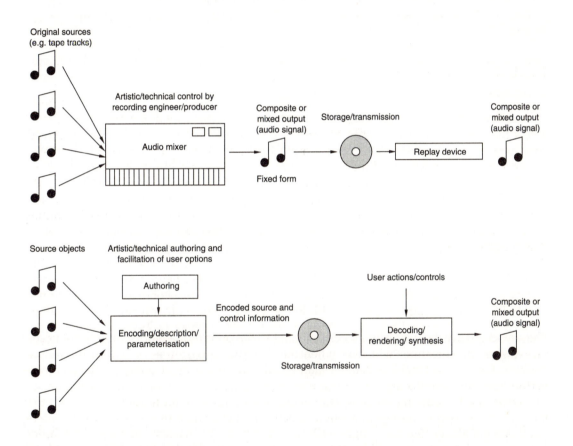

Figure 8.8 (a) In conventional audio production and delivery, sources are combined and delivered at a fixed quality to the user, who simply has to replay it. The quality is limited by the resolution of the delivery link. (b) In some virtual and synthetic approaches the audio information is coded in the form of described objects that are rendered at the replay stage. Here the quality is strongly dependent on the capabilities of the rendering engine and the accuracy of description

consumer's rendering engine, which may involve resynthesis of some elements, based on control data. This is a little like the situation when distributing a song as a MIDI sound file, using General MIDI voices. The audible results, unless one uses downloadable sounds (and even then there is some potential for variation), depends on the method of synthesis and the precise nature of the voices available at the consumer end of the chain.

8.4 Preparing for and understanding release media

Consumer release formats such as CD, DVD, SACD and MP3 usually require some form of mastering and pre-release preparation. This can range from subtle tweaks to the sound quality and relative levels on tracks to PQ encoding, DVD authoring, data encoding and the addition of graphics, video and text. Some of these have already been mentioned in other places in this book.

8.4.1 CD-Audio

PQ encoding for CD mastering can often be done in some of the application packages designed for audio editing, such as SADiE and Sonic Solutions. In this case it may involve little more than marking the starts and ends of the tracks in the play list and allowing the software to work out the relevant frame advances and Red Book requirements for the assembly of the PQ code that will either be written to a CD-R or included in the DDP file for sending to the pressing plant (see Chapter 6). The CD only comes at one resolution and sampling frequency (16 bit, 44.1 kHz) making release preparation a relatively straightforward matter.

8.4.2 DVD

DVD mastering is considerably more complicated than CD and requires advanced authoring software that can deal with all the different options possible on this multi-faceted release format. DVD-Video allows for 48 or 96 kHz sampling frequency and 16, 20 or 24 bit PCM encoding. A two-channel downmix must be available on the disk in linear PCM form (for basic compatibility), but most disks also include Dolby Digital or possibly DTS surround audio. Dolby Digital encoding usually involves the preparation of a file or files containing the compressed data, and a range of settings have to be made during this process, such as the bit rate, dialogue normalisation level, rear channel phase shift and so on. A typical control screen is shown in Figure 8.9. Then of course there are the pictures, but they are not the topic of this book.

There are at least three DVD player types on the market (audio, universal and video), and there are two types of DVD-Audio disc, one containing only audio objects and the other (the DVD-AudioV) capable of holding video objects as well. The video objects on a DVD-AudioV are just the same as DVD-Video objects and therefore can contain video clips, Dolby AC-3 compressed audio and other information. In addition, there is the standard DVD-Video disc, as shown in Figure 8.10.

DVD-AudioV discs should play back in audio players and universal players. Any video objects on an AudioV disk should play back on video-only players. The requirement for

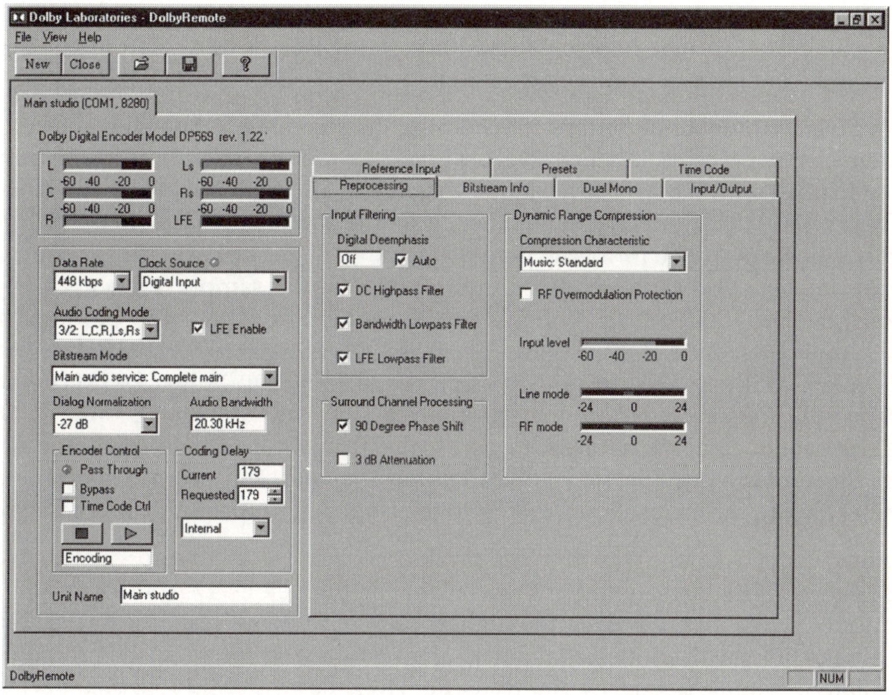

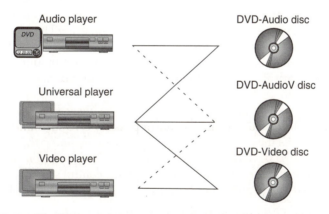

Figure 8.9 Screen display of Dolby Digital encoding software options

Audio player

Universal player

Video player

DVD-Audio disc

DVD-AudioV disc

DVD-Video disc

Figure 8.10 Compatibility of DVD discs and players (Courtesy of DVD working group)

video objects on DVD-AudioV discs to contain PCM audio was dropped at the last moment so that such objects could only contain AC-3 audio if desired. This means that an audio disc could contain a multichannel AC-3 audio stream in a video object, enabling it to be played in a video player. This is a good way of ensuring that a multichannel audio disc plays back in as many different types of player as possible, but requires that the content producer makes sure to include the AC-3 video object in addition to MLP or PCM audio objects. The video object can also contain a DTS audio bitstream if desired.

DVD-Audio has a number of options for choosing the sampling frequencies and resolutions of different channel groups, it being possible to use a different resolution on the front channels from that used on the rear, for example. There are also decisions to be made about the bit budget available on the disk, and whether or not the audio data needs to be MLP encoded for release (see below). The format is more versatile in respect of sampling frequency than DVD-Video, having also accommodated multiples of the CD sample frequency of 44.1 kHz as options (the DVD-Video format allows only for multiples of 48 kHz). Consequently, the allowed sample frequencies for DVD-Audio are 44.1, 48, 88.2, 96, 176.4, 192 kHz. The sample frequencies are split into two groups – multiples of 44.1 and multiples of 48 kHz. While it is possible to split frequencies from one group among the audio channels on a DVD-A (see below), one cannot combine frequencies across the groups for reasons of simple clock rate division. Bit resolution can be 16, 20 or 24 bits per channel, and again this can be divided unequally between the channels, according to the channel group split described below.

Playing time depends on the way in which producers decide to use the space available on the disc, and this requires the juggling of the available bit budget. DVD-Audio can store at least 74 minutes of stereo audio even at the highest sample rate and resolution (192/24). Other modes are possible, with up to six channels of audio playing for at least 74 minutes, using combinations of sample frequency and resolution, together with MLP. Six-channel audio can only operate at the two lower sample rates of either class (44.1/88.2 or 48/96).

A downmixing technique known as SMART (System Managed Audio Resource Technique) is mandatory in DVD-Audio players but optional for content producers. It enables a stereo downmix of the multichannel material to be made in the player but under content producer control, so this information has to be provided at authoring time. The gains, phases and panning of each audio channel can be controlled in the downmix. A separate two-channel mix (L_0/R_0) can be included within an MLP bitstream. If a separate stereo mix is provided on the disc then this is automatically used instead of the player downmix.

All modes other than mono or 2-channel have the option to split the channels into two groups. Group 1 would normally contain the front channels (at least left and right) of the multichannnel balance, while Group 2 could contain the remaining channels. This is known as scalable audio. The resolution of Group 2 channels can be lower than that of Group 1, enabling less important channels to be coded at appropriate resolutions to manage the overall bit budget. The exact point of the split between the channel groups depends on the mode, and there are in fact 21 possible ways of splitting the channels.

It is also possible to 'bit-shift' channels that do not use the full dynamic range of the channel. For example, surround channels that might typically under-record compared with the front channels can be bit shifted upwards so as to occupy only the 16 MSBs of the channel. On replay they are restored to their original gains.

It is not mandatory to use the centre channel on DVD-Audio. Some content producers may prefer to omit a centre speaker feed and rely on the more conventional stereo virtual centre. The merits or demerits of this continue to be debated.

Meridian Lossless Packing (MLP) is licensed through Dolby Laboratories and is a lossless coding technique designed to reduce the data rate of audio signals without compromising sound quality. It has both a variable bit rate mode and a fixed bit rate mode. The variable mode delivers the optimum compression for storing audio in computer data files, but the fixed mode is important for DVD applications where one must be able to guarantee a certain reduction in peak bit rate. The use of MLP on DVD-A discs is optional, but is an important tool in the management of bit budget. Using MLP one would be able to store separate two-channel and multichannel mixes on the same disc, avoiding the need to rely on the semi-automatic downmixing features of DVD players. Owing to the so-called Lossless Matrix technology employed, an artistically controlled L_0/R_0 downmix can be made at the MLP mastering stage, taking up very little extra space on the disc owing to redundancy between the multichannel and two-channel information. MLP is also the key to obtaining high resolution multichannel audio on all channels without scaling.

DVD masters are usually transferred to the pressing plant on DLT tapes, using the Disk Description Protocol (DDP), as described in Chapter 6, or on DVD-R(A) disks as a disk image with a special CMF (cutting master format) header in the disk lead-in area containing the DDP data.

8.4.3 Super Audio CD (SACD)

Version 1.0 of the SACD specification is described in the 'Scarlet Book', available from Philips licensing department. SACD uses DSD (Direct Stream Digital) as a means of representing audio signals, as described in Chapter 2, so requires audio to be sourced in or converted to this form. SACD aims to provide a playing time of at least 74 minutes for both two-channel and six-channel balances. The disc is divided into two regions, one for two-channel audio, the other for multichannel, as shown in Figure 8.11. A lossless data packing method known as Direct Stream Transfer (DST) can be used to achieve roughly 2:1 data reduction of the signal stored on disc so as to enable high quality multichannel audio on the same disc as the two-channel mix.

SACDs can be manufactured as single or dual-layer discs, with the option of the second layer being a Red Book CD layer (the so-called 'hybrid disc'). SACDs, not being a formal part of the DVD hierarchy of standards (although using some of the optical disc technology), do not have the same options for DVD-Video objects as DVD-Audio. The disc is designed first and foremost as a super-high-quality audio medium. Nonetheless there is provision for additional data in a separate area of the disc. The content and capacity of this is not specified but could be video clips, text or graphics, for example. Authoring software enables the text information to be added, as shown in Figure 8.12. SACD masters are normally submitted to the pressing plant on AIT format data tapes (see Chapter 5).

Sony and Philips have paid considerable attention to copy protection and anti-piracy measures on the disc itself. Comprehensive visible and invisible watermarking are standard features of the SACD. Using a process known as PSP (Pit Signal Processing) the width of the pits cut into the disc surface is modulated in such a fashion as to create a visible image on the surface of the CD layer, if desired by the originator. This provides a visible means of authentication. The invisible watermark is a mandatory feature of the SACD layer and is used to authenticate the disc before it will play on an SACD player. The watermark is needed to

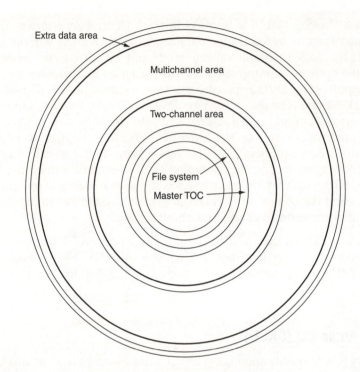

Figure 8.11 Different regions of a Super Audio CD, showing separate two-channel and multichannel regions

decode the data on the disc. Discs without this watermark will simply be rejected by the player. It is apparently not possible to copy this watermark by any known means. Encryption of digital music content is also optional, at the request of software providers.

8.4.4 MP3

MP3, as already explained in Section 2.12, is actually MPEG-1, Layer 3 encoded audio, stored in a data file, usually for distribution to consumers either on the Internet or on other release media. Consumer disk players are increasingly capable of replaying MP3 files from CDs, for example. MP3 mastering requires that the two-channel audio signal is MPEG-encoded, using one of the many MP3 encoders available, possibly with the addition of the ID3 tag described in Chapter 6. Some mastering software now includes MP3 encoding as an option.

Some of the choices to be made in this process concern the data rate and audio bandwidth to be encoded, as this affects the sound quality. The lowest bit rates (e.g. below 64 kbit s^{-1}) will tend to sound noticeably poorer than the higher ones, particularly if full audio bandwidth is retained. For this reason some encoders limit the bandwidth or halve the sampling frequency for very low bit rate encoding, because this tends to minimise the unpleasant side-effects of MPEG encoding. It is also possible to select joint stereo coding mode, as this will improve the technical quality somewhat at low bit rates, possibly at the expense of stereo imaging accuracy. As mentioned above, at very low bit rates some audio processing may be required to make sound quality acceptable when squeezed down such a small pipe.

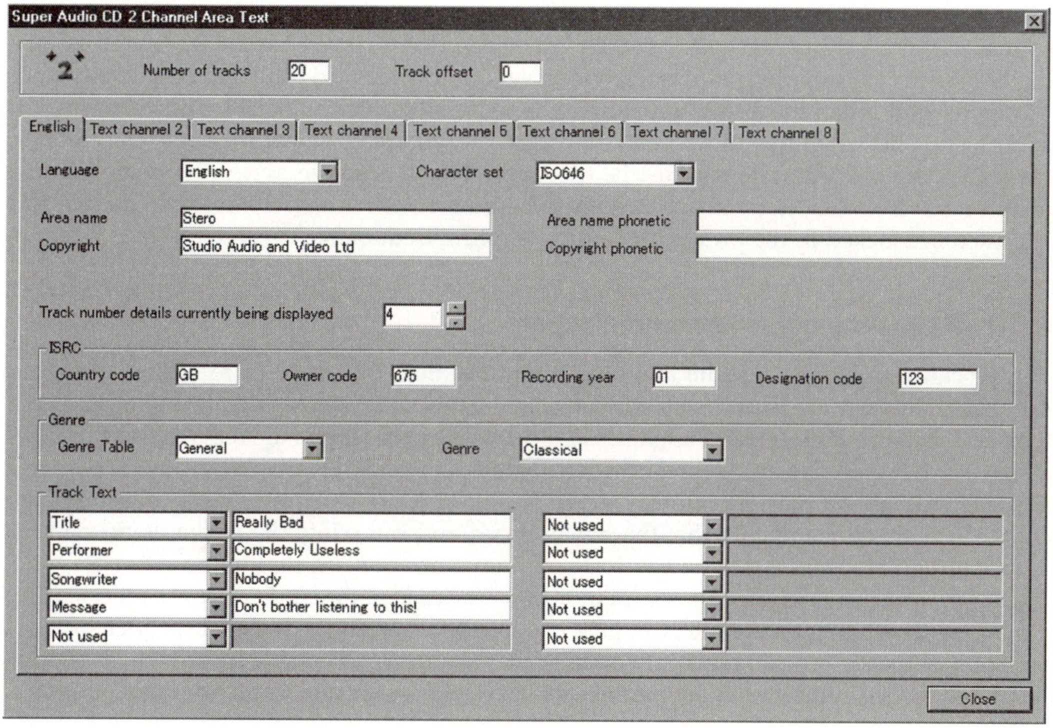

Figure 8.12 Example of SACD text authoring screen from SADiE

8.4.5 MPEG-4, web and interactive authoring

Commercial tools for interactive authoring and MPEG-4 encoding are only just beginning to appear at the time of writing. Such tools enable audio scenes to be described and data encoded in a scalable fashion so that they can be rendered at the consumer replay end of the chain, according to the processing power available.

Interactive authoring for games is usually carried out using low-level programming and tools for assembling the game assets, there being few universal formats or standards in this business at the present time. It requires detailed understanding of the features of the games console in question and these platforms differ considerably. Making the most of the resources available is a specialised task, and a number of books have been written on the subject (see Further reading at the end of this chapter). Multimedia programs involving multiple media elements are often assembled using authoring software such as Director, but that will not be covered further here. Preparing audio for web (Internet) delivery is also a highly specialised topic covered very well in other books (see Further reading).

8.5 Synchronisation

There are many cases in which it is necessary to ensure that the recording and replay of audio and/or MIDI data are time-synchronised to an external reference of some sort. Under the heading of synchronisation comes the subject of locking both recording and replay to a source of SMPTE/EBU timecode or MIDI TimeCode (MTC) (see Chapter 4), as well as locking to an external sampling rate clock, video sync reference or digital audio sync reference. This is needed when the audio workstation is to be integrated with other audio and video equipment, and when operating in an all-digital environment where the sampling frequencies of interconnected systems must be the same. The alternative, when the workstation is operating in isolation, is for all operations to be performed with relation to an internal timing reference locked to the prevailing audio sampling frequency.

8.5.1 Requirements for synchronisation

The synchronisation of an audio application requires that the replay or recording speed and sampling frequency are kept in step with an external timing reference and that there is no long-term drift between this external reference and the passage of time in the replayed audio signal. When lock is required to an external reference there is the possibility that this reference may drift in speed, may have timing jitter or may 'jump' in time (if it is a 'real time' reference such as timecode). Such situations require that the replay speed and sampling rate of the workstation be adjusted regularly and continuously to follow any variations in the timing reference, or to 'flywheel' over them, or even ignore them in some cases (e.g. timecode discontiguities). Speed variations, depending on the rate, can give rise to audible artefacts due to clock jitter, or to the variation of the output sampling rate outside the tolerances acceptable by any other digitally interfaced device in the system, requiring care in system design and implementation.

8.5.2 Timecode synchronisation

The most common synchronisation requirement is for replay to be locked to a source of SMPTE/EBU timecode, because this is used universally as a timing reference in audio and video recording. LTC (longitudinal timecode) is an audio signal that can be recorded on a tape and VITC is contained within lines of a video signal, requiring a suitable reader in the workstation sync interface. A number of desktop workstations that have MIDI features lock to MIDI TimeCode (MTC), which is a representation of SMPTE/EBU timecode in the form of MIDI messages. Details of both types of timecode were given in Chapter 4.

It is important to know what kind of synchronisation is used by your hardware and software. One of the factors that must be considered is whether external timecode is simply used as a timing reference against which sound file replay is triggered, or whether the system continues to slave to external timecode for the duration of replay. In some cases these modes are switchable because they both have their uses. In the first case replay is simply 'fired off' when a particular timecode is registered, and in such a mode no long-term relationship is maintained between the timecode and the replayed audio. This may be satisfactory for some basic operations but is likely to result in a gradual drift between audio replay and the external reference if files longer than a few seconds are involved. It may be useful though, because replay remains locked to the workstation's internal clock reference, which may be more stable than external references, potentially leading to higher audio quality from the system's convertors. Some cheaper systems do not 'clean up' external clock signals very well before using them as the sample clock for D/A conversion, and this can seriously affect audio quality.

In the second mode a continuous relationship is set up between timecode and audio replay, such that long-term lock is achieved and no drift is encountered. This is more difficult to achieve because it involves the continual comparison of timecode to the system's internal timing references and requires that the system follows any drift or jump in the timecode. Jitter in the external timecode is very common, especially if this timecode derives from a video tape recorder, and this should be minimised in any sample clock signals derived from the external reference. This is normally achieved by the use of a high-quality phase-locked loop, often in two stages. Wow and flutter in the external timecode can be smoothed out using suitable time constants in the software that converts timecode to sample address codes, such that short-term changes in speed are not always reflected in the audio output but longer-term drifts are.

Sample frequency conversion may be employed at the digital audio outputs of a system to ensure that changes in the internal sample rate caused by synchronisation action are not reflected in the output sampling rate. This may be required if the system is to be interfaced digitally to other equipment in an all-digital studio.

8.5.3 Synchronisation to external audio, film or video references

In all-digital systems it is necessary for there to be a fixed sampling frequency, to which all devices in the system lock. This is so that digital audio from one device can be transferred directly to others without conversion or loss of quality. In systems involving video it is often necessary for the digital audio sampling frequency to be locked to the video frame rate and for timecode to be locked to this as well. The reference signal is likely to be a 'house sync' composite video signal that does not necessarily carry time-of-day information. It would be used to lock the internal sampling frequency clock of the workstation. An alternative to this is a digital audio sync signal such as word clock or an AES11 standard sync reference (a stable AES3 format signal, without any audio).

Other sync signals could include tachometer or control track pulses from tape machines or frame rate pulses from film equipment. If a system is to be able to resolve to any or all of these, as well as to timecode and digital audio inputs, a very versatile 'gearbox' will be required to perform the relevant multiplications and divisions of synchronisation signals at different rates, such that they can be used to derive the internal sampling rate clock of the system. A stable voltage-controlled oscillator (VCO) and phase-locked loop are commonly used for this purpose.

Figure 8.13 (page 260) illustrates a possible conceptual diagram of synchronised operation, with a variety of references and a constant sampling rate output. The sampling frequency convertor is not necessary if suitably constant external relationships can be maintained between the different forms of sync signal and the audio sampling frequency.

8.6 System troubleshooting

8.6.1 Troubleshooting MIDI

When a MIDI system fails to perform as expected, or when devices appear not to be responding to data being transmitted from a controller, it is important to adopt logical fault-finding

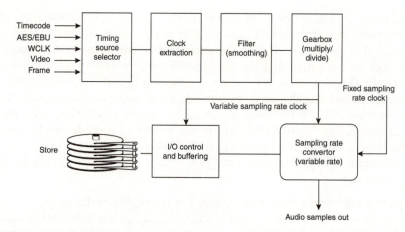

Figure 8.13 Conceptual diagram of replay synchronised to one of a number of timing sources. Blocks of data are fetched from disk at a rate determined by the current sampling clock

techniques rather than pressing every button in sight and starting to replug cables. The fault will normally be a simple one and there are only a limited number of possible causes. It is often worth starting at the end of the system nearest to the device that exhibits the problem and working backwards towards the controller, asking a number of questions as you go. You are basically trying to find out either where the control signal is getting lost or why the device is responding in a strange way.

Look at the hints in Figure 8.14. Firstly, is MIDI data getting to the device in question? Most devices have some means of indicating that they are receiving MIDI data, either by a flashing light on the front panel or some other form of display. Alternatively it is possible to buy small analysers which in their simplest form may do something like flashing a light if MIDI data is received. If data is getting to the device then the problem is probably either within the device or after its audio output. The most common mistake that people make is to think that they have a MIDI problem when in fact they have an audio problem. Check that the audio output is actually connected to something and that its destination is turned on and faded up. Plug in a pair of headphones to check if the device is responding to MIDI data. If sound comes out of the headphones then the problem most probably lies in the audio system.

If the device is receiving MIDI data but not producing an audio output, try setting the receive mode to 'omni on' so that it responds on all channels. If this works then the problem must be related to the way in which a particular channel's data is being handled. Check that the device is enabled to receive on the MIDI channel in question. Check that the volume is set to something other than zero and that any external MIDI controllers assigned to volume are not forcing the volume to zero (such as any virtual faders in the sequencer package). Check that the voice assigned to the channel in question is actually assigned to an audio output that is connected to the outside world. Check that the main audio output control on the unit itself is turned up. Also try sending note messages for a number of different notes – it may be that the voice in question is not set up to respond over the whole note range.

If no MIDI data is reaching the device then move one step further back down the MIDI signal chain. Check the MIDI cable. Swap it for another one. If the device is connected to a MIDI

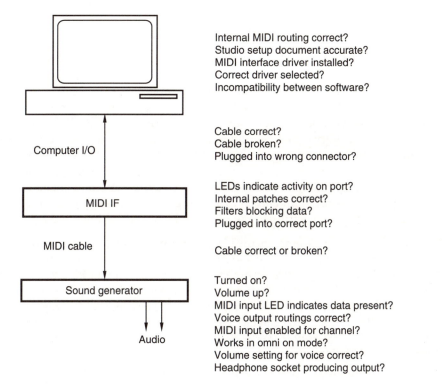

Internal MIDI routing correct?
Studio setup document accurate?
MIDI interface driver installed?
Correct driver selected?
Incompatibility between software?

Cable correct?
Cable broken?
Plugged into wrong connector?

LEDs indicate activity on port?
Internal patches correct?
Filters blocking data?
Plugged into correct port?

Cable correct or broken?

Turned on?
Volume up?
MIDI input LED indicates data present?
Voice output routings correct?
MIDI input enabled for channel?
Works in omni on mode?
Volume setting for voice correct?
Headphone socket producing output?

Figure 8.14 A number of suggestions to be considered when troubleshooting a MIDI system

router of some kind, check that the router input receiving the required MIDI data is routed to the output concerned. Try connecting a MIDI keyboard directly to the input concerned to see if the patch is working. If this works then the problem lies further up the chain, either in the MIDI interface attached to the controller or in the controller itself. If the controller is a computer with an external MIDI interface, it may be possible to test the MIDI port concerned. The setup software for the MIDI interface may allow you to enter a 'Test' mode in which you can send unspecified note data directly to the physical port concerned. This should test whether or not the MIDI interface is working. Most interfaces have lights to show when a particular port is receiving or transmitting data, and this can be used for test purposes. It may be that the interface needs to be reconfigured to match a changed studio setup. Now go back to the controller and make sure that you are sending data to the right output on the required MIDI channel and that you are satisfied, from what you know about it, that the software concerned should be transmitting.

If no data is getting from the computer to the interface, check the cables to the interface. Then try resetting the interface and the computer. This sometimes re-establishes communication between the two. Reset the interface first, then the computer, so that the computer 'sees' the interface (this may involve powering down, then up). Alternatively, a soft reset may be possible using the setup software for the interface. If this does not work, check that no applications are open on the computer which might be taking over the interface ports concerned (some applications will not give up control over particular I/O ports easily). Check the configuration of any software MIDI routers within the computer to make sure that MIDI data is 'connected' from the controlling package to the I/O port in question.

Ask yourself the question: 'Was it working the last time I tried it?' If it was, it is unlikely that the problem is due to more fundamental reasons such as the wrong port drivers being installed in the system or a specific incompatibility between hardware and software, but it is worth thinking through what you have done to the system configuration since the last time it was used. It is possible that new software extensions or new applications may conflict with your previously working configuration, and removing them will solve the problem. Try using a different software package to control the device which is not responding. If this works then the problem is clearly with the original package. Assuming that the device in question had been responding correctly on a previous occasion, any change in response to MIDI messages such as program and control changes is most likely due either to an altered internal setup or a message getting to the device which was not intended for it.

Most of the internal setup parameters on a MIDI-controlled device are accessible either using the front panel or using system exclusive messages. It is often quite a long-winded process to get to the parameter in question using the limited front panel displays of many devices, but it may be necessary to do this in order to check the intended response to particular MIDI data. If the problem is one with unusual responses (or no response) to program change messages then it may be that the program change map has been altered and that a different stored voice or patch is being selected from the one intended. Perhaps the program change number in question is not assigned to a stored voice or patch at all. If the device is switching between programs when it should not then it may be that your MIDI routeing is at fault. Perhaps the device is receiving program changes intended for another. Check the configuration of your MIDI patcher or multiport interface. A similar process applies to controller messages. Check the internal mapping of controller messages to parameters, and check the external MIDI routing to make sure that devices are receiving only the information intended for them.

When more than one person uses a MIDI-controlled studio, or when you have a lot of different setups yourself, virtually the only way to ensure that you can reset the studio quickly to a particular state is to store system exclusive dumps of the full configuration of each device and to store any patcher or MIDI operating system maps. These can either be kept in separate librarian files or as part of a sequence, to be downloaded to the devices before starting the session. Once you have set up a configuration of a device that works for a particular purpose it should be stored on the computer so that it could be dumped back down again at a later date.

8.6.2 Digital interface troubleshooting

If a digital interface between two devices appears not to be working it could be due to one or more of the following conditions. These are covered in more detail in *The Digital Interface Handbook* (see Further reading).

Asynchronous sample rates

The two devices must normally operate at the same sampling frequency, preferably locked to a common reference. Ensure that the receiver is in external sync mode and that a synchronizing signal (common to the transmitter) is present at the receiver's sync input. If the incoming signal's transmitter cannot be locked to the reference it must be resynchronized or sample rate converted. Alternatively, set the receiver to lock to the clock contained in the digital audio input (standard two-channel interfaces only).

'Sync' or 'locked' indicator flashing or out on the receiver normally means that no sync reference exists or that it is different from that of the signal at the digital input. Check that sync reference and input are at correct rate and locked to the same source. Decide on whether to use internal or external sync reference, depending on application.

If problems with 'good lock' or drifting offset arise when locking to other machines or when editing, check that any timecode is synchronous with video and sampling rate.

Sampling frequency mode

The transmitter may be operating in the AES3 single-channel-double-sampling-frequency mode in which case successive sub-frames will carry adjacent samples of a single channel at twice the normal sampling frequency. This might sound like audio pitch-shifted downwards if decoded and converted by a standard receiver incapable of recognising this mode. Alternatively the devices may be operating at entirely different sampling frequencies and therefore not communicate.

Digital input

It may be that the receiver is not switched to accept a digital input.

Data format

Received data is in the wrong format. Both transmitter and receiver must operate to the same format. Conflicts may exist in such areas as channel status, and there may be a consumer–professional conflict. Use a format convertor to set the necessary flags.

Non-audio or 'other uses' set

The data transmitted over the interface may be data-reduced audio, such as AC-3 or DTS format. It can only be decoded by receivers specially designed for the task. The data will sound like noise if it is decoded and converted by a standard linear PCM receiver, but in such receivers it will normally be muted because of the indication in channel status and/or the validity bit.

Cables and connectors

Cables or connectors may be damaged or incorrectly wired. Cable may be too long, of the wrong impedance, or generally of poor quality. Digital signal may be of poor quality. Check eye height on scope against specification and check for possible noise and interference sources. Alternatively make use of an interface analyser.

SCMS (consumer interface only)

Copy protect or SCMS flag may be set by transmitter. For professional purposes, use a format convertor to set the necessary flags or use the professional interface which is not subject to SCMS.

Receiver mode

Receiver is not in record or input monitor mode. Some recorders must be at least in record–pause before they will give an audible and metered output derived from a digital input.

8.6.3 Troubleshooting software

This could form a book in its own right and depends a lot on the operating system and applications in question. There are, however, a few rules of thumb to be observed when trying to get software to work.

Firstly, make sure you have all the latest updates and revisions to the current system software and applications. Latest versions tend to be reasonably safe together. Patches and updates can often be downloaded from the Internet. Check also that the memory and CPU requirements of the application are met. Begin with a basic set of system extensions and don't load any more software or extensions than you need onto an audio workstation. General purpose extensions and third-party software can sometimes conflict with the smooth operation of audio workstation packages and many people run only audio software on such platforms rather than trying to use them as general purpose computers as well.

Make sure that you are using the correct and latest drivers for any sound cards and MIDI interfaces in the system and that the disk interface and drivers are suitable for high speed audio and video operation.

Further reading

Beggs, J. and Thede, D. (2001) *Designing Web Audio*. O'Reilly and Associates.
Boer, J. (2002) *Game Audio Programming*. Charles River Media.
Marks, A. (2001) *The Complete Guide to Game Audio*. CMP Books.
Rumsey, F. (2001) *Spatial Audio*. Focal Press.
Rumsey, F. and Watkinson, J. (2004) *The Digital Interface Handbook*, third edition. Focal Press.

Index

Focal Press

www.focalpress.com
Join Focal Press on-line
As a member you will enjoy the following benefits:

- an email bulletin with **information on new books**

- a regular **Focal Press Newsletter**:
    - featuring a selection of new titles
    - keeps you informed of **special offers, discounts and freebies**
    - alerts you to **Focal Press news and events** such as author signings and seminars

- complete access to **free content** and reference material on the focalpress site, such as the focalXtra articles and commentary from our authors

- a **Sneak Preview** of selected titles (sample chapters) *before* they publish

- a chance to have your say on our **discussion boards** and **review books** for other Focal readers

Focal Club Members are invited to give us feedback on our products and services.
Email: worldmarketing@focalpress.com – we want to hear your views!

Membership is **FREE**. To join, visit our website and register. If you require any further information regarding the on-line club please contact:

Lucy Lomas-Walker
Email: l.lomas@elsevier.com
Tel: +44 (0) 1865 314438
Fax: +44 (0)1865 314572
Address: Focal Press, Linacre House,
Jordan Hill, Oxford, UK, OX2 8DP

Catalogue
For information on all Focal Press titles, our full catalogue is available online at www.focalpress.com and all titles can be purchased here via secure online ordering, or contact us for a free printed version:

USA
Email: christine.degon@bhusa.com
Tel: +1 781 904 2607 T

Europe and rest of world
Email: j.blackford@elsevier.com
Tel: +44 (0)1865 314220

Potential authors
If you have an idea for a book, please get in touch:

USA
editors@focalpress.com

Europe and rest of world
focal.press@elsevier.com